Writing in the Biological Sciences

A COMPREHENSIVE RESOURCE FOR
SCIENTIFIC COMMUNICATION

Angelika H. Hofmann, Ph.D.
Yale University

New York   Oxford
OXFORD UNIVERSITY PRESS

This book is dedicated to the memory of
Francisco Javier Triana Alonso,
who was and always will be larger than life to me

Oxford University Press is a department of the University of Oxford. It furthers the University's objective of excellence in research, scholarship, and education by publishing worldwide.

Oxford   New York
Auckland   Cape Town   Dar es Salaam   Hong Kong   Karachi
Kuala Lumpur   Madrid   Melbourne   Mexico City   Nairobi
New Delhi   Shanghai   Taipei   Toronto

With offices in
Argentina   Austria   Brazil   Chile   Czech Republic   France   Greece
Guatemala   Hungary   Italy   Japan   Poland   Portugal   Singapore
South Korea   Switzerland   Thailand   Turkey   Ukraine   Vietnam

Oxford is a registered trademark of Oxford University Press
in the UK and certain other countries.

Published in the United States of America by
Oxford University Press
198 Madison Avenue, New York, NY 10016

© Oxford University Press 2013

ISBN: 978-0-19-976528-7

Printing Number: 9 8 7 6 5 4 3 2 1

Printed in the United States of America
on acid-free paper

BRIEF CONTENTS

CONTENTS

v

FOREWORD

Academic science has a language of its own. Written and spoken by its practitioners, this language is sometimes practically unintelligible to the uninitiated. As a college or graduate-level student beginning to write scientific literature with highly technical content—papers, abstracts, proposals—your biggest challenge is to go beyond learning the jargon to convey complex factual information clearly and logically. As an instructor, your challenge is to teach your students this language through classroom practices. As a researcher, your challenge is to be fluent in the scientific language, producing peer-reviewed grants, manuscripts, and talks that may even reshape the language itself.

Despite years of training, in my experience, most new faculty researchers in science and medical fields are shocked and beleaguered by the writing workload required to build and sustain an independent research group. While they are scientifically ready for their new roles, much less attention has been paid to preparing their communication abilities. Yet, successful faculty researchers are also those with the communication experience and ability to advocate for their science effectively—especially as the pool of funds available for research continues to decline. To help prepare trainees for the challenges that lie ahead, the importance of scientific communication must be emphasized during college and graduate school. Moreover, as students practiced in scientific writing and communication also develop the ability to think logically and critically about the subject matter, these skills can be viewed as tools with which to train the budding scientific mind. Thus, the principles of sound scientific writing should be viewed as foundational and should be integrated throughout the curriculum of higher education much as chemistry is for those majoring in biology.

Why is the craft of scientific communication so often pushed to the wayside? This essential discipline requires specific educational resources to guide instructor and student. This new text, *Writing in the Biological Sciences: A Comprehensive Guide to Scientific Communication*, is an invaluable contribution to the field. Through an accessible, clear, concise, yet thorough treatment of the subject, it instructs in the unique language of scientific communication. The book's practical approach of breaking down the most relevant forms of writing and presentation into their component parts enables the student to internalize key common principles that are discussed throughout the text. The text also demonstrates that those adept in scientific communication in the classroom—be it through lab reports, term papers, or in-class oral presentations—are those well on their way to crafting independent research publications, review articles, and seminar presentations.

Much like a box of paints, the techniques and information in this book are your toolkit with which you will communicate your own ideas and contributions to the outside world. You may be opening this book for one of many reasons. Perhaps you are using it as a required textbook. Perhaps you are coming to the formalized study of scientific writing for the first time, or perhaps you are someone who is seeking specifically to improve your scientific writing skill. Keep in mind that as with a foreign language, true proficiency in scientific communication can take years to achieve. Hold on to this volume in the years to come, because you will refer to the guidelines within again and again.

Tammy Wu, Ph.D.
Yale University

PREFACE

Communication plays a fundamental role in the sciences. It is the engine that propels virtually all scientific progress. Without good communication skills, scientists stand little chance of publishing their work, obtaining funds, attracting a wide audience when giving a talk, or getting a new job. Even the most promising discovery means little if it cannot be communicated successfully. In fact, it is more often the case that advancements are limited not by their technical merit but by ineffective articulation. Thus, clear communication is a requirement, not an option, for a good scientist.

Writing in the Biological Sciences: A Comprehensive Resource for Scientific Communication serves as an all-inclusive "one-stop" reference guide to scientific writing and communication for budding professionals in the life sciences and related fields. The book is intended as a free-standing textbook for a corresponding course on writing in the sciences as well as an accompanying text or reference guide in courses with writing-intensive components. It covers all the basics of scientific communication that students need to know and master for successful scientific careers. The book lays the foundations for future professional writing by starting with basic scientific writing principles and applying these to composing lab reports, summaries, and critiques and eventually to full-fledged scientific research articles, review articles, and grant proposals. Practical advice for organizing academic presentations and posters as well as for putting together job applications is also included.

Through extensive and annotated practical examples and easy-to-understand and easy-to-remember principles and guidelines, this book explains how to write clearly as a scientific author and how to recognize shortcomings in one's own writing, as well as that of others. It does so not only by providing crucial knowledge about the structure and delivery of written material but also by explaining how readers go about reading. Furthermore, potential problem areas in written and oral presentations are pointed out, and many examples are provided for wording certain sections in research papers, grant proposals, or scientific talks.

There are numerous hallmark features of this text, including:

Practical organization. As a result of extensive class testing, the text has been revised repeatedly to reflect the interests, concerns, and problems undergraduate students encounter when learning how to communicate in their disciplines. The table of contents is divided into four distinctive parts. Parts I and II follow a logical progression from the basics of scientific writing style and composition to constructing effective figures and tables. Part III applies these basic principles to planning and organizing foundational precursors of scientific publications, namely lab reports, summaries, and critiques, to the specifics of how to write each major section of a scientific research paper and review article. Part IV covers more advanced scientific communication, including how to compose grant proposals, posters, oral presentations, and job applications.

Comprehensive coverage. The text includes detailed discussions of the main scientific documents undergraduate students in the sciences encounter, including laboratory reports, scientific research papers, review articles, proposals, oral presentations, posters, and job applications. The broad coverage of multiple forms of communication allows the text to serve as a writing guide for science writing courses as well as a companion text for advanced biology courses that have a writing-intensive component. In addition, the text provides comprehensive coverage of writing mechanics, style, and composition.

Numerous real-world and relevant examples. The in-chapter examples are derived directly from real lab reports, scientific research papers, and grant proposals in the biological sciences. Throughout the chapters, common pitfall examples are followed by successful revisions, and annotated examples provide explanations of various text elements and concepts. These annotated examples of text passages bring to life the guidelines and principles presented throughout the chapters.

Extensive exercise sets and end-of-chapter summaries. Chapter summaries reference the most important concepts in an easy-to-understand bulleted format. The end-of-chapter exercises and problems review style and composition principles and encourage students to apply the presented principles and guidelines to their own writing. Answers to the exercises are provided in a separate appendix.

Writing guidelines and checklists for revisions. Straightforward writing principles and guidelines presented in the book provide the basis for writing scientific articles, proposals, and job applications and for creating clear posters and oral presentations. Explanations of these writing principles and guidelines are followed by common pitfall examples, as well as by suggestions and advice to revise one's work successfully. Checklists at the end of each chapter aid readers in remembering and applying these rules when writing or revising a document.

Sample wording for scientific documents and presentations. Beginning scientific writers will especially value the many tables with sample sentences that apply to different sections of a scientific research article, review article, or grant proposal. Anyone presenting data at meetings and conferences will find the sample phrases and advice on creating and delivering a talk or poster highly useful.

Writing in the Biological Sciences: A Comprehensive Resource for Scientific Communication teaches students how to practice writing and thinking in the sciences and lets them learn by example from the writings of others. Familiarity with the nuances of these elements will be enhanced as students read scientific literature and pay attention to how professional scientists write about their work. Students will see improvement in their own writing skills by repeatedly practicing reading, writing, and critiquing of others' work. Writing a clear research article or grant proposal or presenting an articulate talk can be difficult for any scientist, but this difficulty is by no means insurmountable. Ultimately, with guidance and practice, students should be able to write papers or proposals that sparkle with clarity and deliver engaging presentations. As they write their own papers or prepare their own posters, they will recognize that every project has its unique challenges and that they will need practice and good judgment to apply all the writing and communication principles presented herein. In giving due attention to composition, style, and impact, students will improve their communication skills significantly, and this book will have accomplished its purpose.

ACKNOWLEDGMENTS

This book includes many ideas and "specimens" that I, as a scientist, instructor, and editor have collected over the years. A few of these "specimens" are originals. Many are a variation of someone else's original. Others are cited intact from their respective sources. Without these "specimens" and samples, this book would not have been possible. For their contribution, I would therefore like to especially acknowledge my students, friends, and colleagues from the Max-Planck Institute, Fritz-Haber Institute, Humboldt University, Karolinska Institute, University of Carabobo, University of Massachusetts at Worcester, and Yale University who have shared information and ideas across the sciences. I am particularly thankful to all those who were courageous enough to allow me to use draft sentences, paragraphs, or sections as examples or problems in this book as well as to those providing me with extensive and very specific samples: Irene Bosch, Mark Bradford, Jaclyn Brown, Stephane Budel, Philip Duffy, Mónica I. Feliú-Mójer, Nikolas Franceschi Hofmann, Alison Galvani, Roland Geerken, Jun Korenaga, Amanda Miller, Klaus von Schwarzenberg, Orli Steinberg, and Tammy Wu. Without these samples the book would not be nearly as effective in exemplifying clear writing.

I am also grateful to all the reviewers who have edited and commented on various draft chapters, including

Daniel G. Blackburn
Trinity College

Chad E. Brassil
University of Nebraska–Lincoln

Heather Bruns
Ball State University

Arthur Buikema
Virginia Tech

Alyssa C. Bumbaugh
Shippensburg University of Pennsylvania

Douglas J. Burks
Wilmington College of Ohio

Lynn L. Carpenter
University of Michigan

Dale Casamatta
University of North Florida

Deborah Cato
Wheaton College

David T. Champlin
University of Southern Maine

Kendra Cipollini
Wilmington College

Francisco Cruz
Georgia State University

Charles Elzinga
Michigan State University

Kirsten Fisher
California State University, Los Angeles

Laurel Fox
University of California, Santa Cruz

Blaine Griffen
University of South Carolina

Glenn Harris
Virginia State University

Leif Hembre
Hamline University

Evelyn N. Hiatt
Kentucky Wesleyan College

W. Wyatt Hoback
University of Nebraska at Kearney

Terry Keiser
Ohio Northern University

Tali Lee
University of Wisconsin, Eau Claire

Joshua Mackie
San Jose State University

Nusrat Malik
University of Houston

Jesse M. Meik
The University of Texas at Arlington

Jennifer A. Metzler
Ball State University

Daniel Moon
University of North Florida

Barbara Musolf
Clayton State University

Michael O'Donnell
Trinity College

T. Page Owen, Jr.
Connecticut College

Dr. Helen Piontkivska
Kent State University

Ann E. Rushing
Baylor University

Robert S. Stelzer
University of Wisconsin, Oshkosh

Ken G. Sweat
Arizona State University, West Campus

Katerina Thompson
University of Maryland, College Park

Neal J. Voelz
St. Cloud State University

Particular thanks go to Betty Liu, Fiona Bradford, Paola Crucitti, Riccardo Missich, Rudolf Lurz, Tammy Wu, Francois Franceschi, Jennifer Powell, Stephane Budel, and John Alvaro for their encouragement as well for as their critical comments and the many helpful discussions over the years. My deepest gratitude is in memory of Francisco Triana Alonso for his unwavering belief, support, and pride in me and this endeavor—beginning with my first sprouting ideas, to the delivery of them to students, and ultimately, to the publication of this book.

Finally, I would like to express appreciation to everyone at Oxford University Press: Jason Noe, senior editor; Caitlin Kleinschmidt, editorial assistant; Patrick Lynch, editorial director; John Challice, publisher and vice president; Jason Kramer, marketing manager; Frank Mortimer, director of marketing; Lisa Grzan, production team leader; David Bradley, production editor; and Michele Laseau, art director.

PART ONE

Scientific Writing Basics

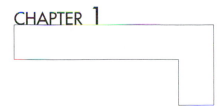

CHAPTER 1

Scientific Communication

1.1 IMPORTANCE OF SCIENTIFIC COMMUNICATION

Communication plays a tremendously important role in the sciences. It is the engine that propels virtually all scientific progress. With good communication skills, scientists stand a much better chance of publishing their work, obtaining funds, or attracting a wide audience. Even the most promising discovery means little if it cannot be communicated successfully. Thus, to be a good scientist, clear communication is a requirement, not an option.

To write clearly in the sciences, you have to use professional scientific writing tools. Such tools include basic writing principles of style, composition, and format. Most importantly, you need to understand what readers expect to find and how they go about reading. Expectations and perceptions of readers have been widely studied in the fields of rhetoric, linguistics, and cognitive psychology. This book makes use of these findings, explains them, and teaches you how to apply them to a broad range of scientific documents, including lab reports, scientific research papers, term papers, review articles, and research proposals. Learning how to compose such documents well lays the foundation for your future writing of professional scientific texts, which include:

- Scientific research articles—to communicate findings to other scientists and to the public
- Review articles—to glean and communicate in depth interpretations of current topics
- Grant proposals—to apply for funding of research

Skills learned in preparing posters, oral presentations, and job application as an undergraduate will equally be of benefit later in your career path.

This book also aims to lay the foundation for each of these scientific communication avenues by providing you with important fundamentals and step-by-step guidelines.

1.2 READERS AND WRITERS

In science, readers are affected by the content and format of a paper as well as by its composition and style. Their interpretations are based not only on words, sentences, and paragraphs but above all on the structural location of these elements. In fact, the logical and structural organization of your document is much more important than using perfect grammatical form (see Figure 1.1).

As an author, you need to be conscious of these elements. Understanding the correlation of structure and function in a sentence, paragraph, or section is what underlies the science of scientific writing.

The main reason for bad style and composition in scientific writing is lack of training. The majority of scientific writers are unaware of how to identify words, sentences, or paragraphs that may give readers a problem. Most are aware of certain "rules" taught to them in high school English composition classes but not of the fine-tuned principles that would benefit them as professionals in a scientific field.

To be a successful scientist and professional writer, basic English composition is usually not enough. You need to know what information readers in the field expect to find and where. You also need to realize that all too often you are too close to your own writing to judge it. The principles and guidelines presented in the earlier chapters of this book allow you to lay a solid foundation and to start obtaining the tools of professional writing. The later chapters of the book let you apply and expand on these tools as you move closer and closer into the professional world.

The book also provides you with step-by-step, concrete examples and exercises to help you to grasp fully the application and use of the principles within the necessary format expected for the respective documents. However, reading and hearing about these principles

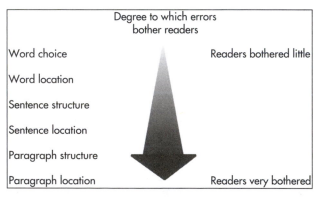

Figure 1.1 Degree to which errors in writing bother readers.

and guidelines is not enough. You, as the writer, must practice writing and thinking within this structure and must learn by example from the writings of others. Familiarity with the nuances of these elements will be enhanced as you read scientific literature and pay attention to how professional scientists write about their work. In giving due attention to composition, style, and impact, your communication skills will evolve into professional writing, and this book will have accomplished its purpose.

1.3   NECESSARY BASIC SKILLS

To compose scientific documents and presentations, technical expertise is needed. This expertise includes computer literacy and skills in specific computer programs, including:

- MS Word or Pages for writing and formatting documents
- Excel or Numbers for making tables and preparing graphs
- PowerPoint or Keynote for creating slides and posters for presentations

If you find yourself unfamiliar with any of these programs, consider taking some classes or completing an online tutorial (see also Sections 5.5 and 12.5). It is essential to know these basic skill sets.

1.4   ABOUT THIS BOOK

This book is both a practical text with teaching exercises and a ready reference guide that can be used long after class is over. In writing this book, my goal was to target as broad of an undergraduate scientific audience as possible, ranging from lower-level to upper-level and honors students. The overall objective was to lay the foundations for future professional scientific writing. Thus, for example, lab reports that follow the general format provided in this book provide the groundwork for later scientific research articles that get published in academic journal; term papers are the basis for review articles; and research proposals for grant proposals.

To set the stage for future professional writing, initial chapters of the book discuss the basics of scientific writing. You may already know some of the basic principles presented in these beginning chapters, other principles you may have forgotten or may not have learned yet. The point is for these chapters to lay a solid foundation of scientific writing principles that you can refer to throughout your university time and beyond. Subsequent chapters provide the basics of displaying and analyzing data, lower-level undergraduate writing, and upper-level undergraduate documents and presentations.

In each chapter, specific, easy-to-remember guidelines are highlighted. For each principle or guideline, good—and sometimes also bad—examples are provided. Many of these examples have annotations. End-of-chapter exercises encourage you to practice applying the stated principles and guidelines. The way you will learn and improve is by trying to apply the relevant writing principles yourself—even if you miss the point in your answers. It is important to practice on paper because the only way to learn how to write is by writing. You may disagree with some choices of answers. In fact, many exercises may have more than one possible answer, as there may be more than one way to express an idea and to improve written passages. Some choices are better, some worse, and some just a matter of personal style. The goal is to improve the original statements such that they are clearer and more easily understandable by the reader.

CHAPTER 2

Literature Sources

2.1   SEARCHING THE LITERATURE

Reading and understanding scientific literature, writing laboratory reports, composing a research paper, or preparing a review article or thesis typically requires you to be able to search for appropriate information, especially online. Using such information allows you to apply up-to-date research in your laboratory reports and essays that would not normally be available in textbooks. Also, if you submit coursework that includes references and information from relevant and recent publications it shows that you have made an effort to research your work thoroughly.

Several sources of information exist. These include books, journals, newspapers, and the web. Because of the time it takes to publish a book, books usually contain more dated information than will be found in the most recent journals and newspapers. A huge amount of information is available online, but it may not be immediately clear what resource to use for the purpose of composing a scientific document. Internet search engines may not prove ideal for obtaining information to compose a scientific document, because you may find it hard to distinguish relevant, reliable, and authoritative information and to obtain full–length publications. A few internet sites, such as Google Scholar (http://scholar.google.com/schhp?hl=en&as_sdt=0,11), provide access to primary literature, that is, peer-reviewed research articles. Generally, however, bibliographic databases such as Web of Science and your library offer a more reliable and structured way of finding good-quality information. Such databases can provide links to the full text of scholarly journals, in which you find the most up-to-date information on research in industry and academia.

Unlike most information found online, work published in scholarly journals is typically peer-reviewed, meaning that is has undergone a rigorous evaluation process by experts in the field to maintain standards and provide credibility. Scientific journal articles are usually considered primary sources because they usually report results for the first time (in contrast to secondary sources, which analyze and discuss the information provided by primary sources, and tertiary sources, such as textbooks, which compile and reorganize information provided in mainly secondary sources).

Potentially relevant scientific information available in scholarly journals can be gleaned by searching online databases such as MEDLINE®, SCOPUS®, BIOSIS, and the Web of Science using, for example, keywords, author, source, author's affiliation, cited author, cited work, and cited year.

Top science databases include:

PubMed	PubMed comprises more than 20 million citations for biomedical literature from MEDLINE, life science journals, and online books. Citations may include links to full-text content from PubMed Central and publisher web sites.
MEDLINE®	MEDLINE, produced by the National Library of Medicine, part of the National Institutes of Health, covers more than 4000 journal titles and is international in scope. Broad coverage includes basic biomedical research and clinical sciences.
SCOPUS™	Scopus provides broad international coverage of the sciences and social sciences, indexing 14,000 journals.
BIOSIS	Biosis, the online version of Biological Abstracts and Biological Abstracts-Reports, Reviews, Meetings contains literature references from all of the life sciences. This is the premier database for coverage of botany research.
Web of Science	The ISI Citation Databases collectively index more than 8000 peer-reviewed journals. It provides web access to Science Citation Index Expanded, which covers 6300 international science and engineering journals.
Current Contents®	Current Contents Science Edition covers all the Science editions of the Current Contents Search® database in one package.

Note that not all libraries may have access to all databases. Alternate, freely accessible web search engines include:

Google Scholar	This site indexes the full text of most peer-reviewed online journals of Europe and America's largest scholarly publishers.
Scirus	Scirus is a comprehensive science-specific search engine from Elsevier, but not all of its results include full text. Scirus shares its scientific search results with SCOPUS.
CiteSeer^x	CiteSeerX indexes scientific articles with a focus on computer and information science. It currently has over 1.5 million documents.

| getCITED | getCITED makes over 3,000,000 publications of journal articles, book chapters, and other publications publically available. Its information is entered by members and includes both peer-reviewed and nonreviewed publications. |

Note that in many cases search results link directly to commercial journal articles and that a majority of the time users will only be able to access a brief summary of the articles. You may have to pay a fee to access entire articles. A longer list of major databases and search engines for finding and accessing articles in academic journals can be found at http://en.wikipedia.org/wiki/Academic_databases_and_search_engines.

2.2   SELECTING REFERENCES

If you already know the details of a journal article, you can use the journal search to check if the library holds the article by searching for the title of the journal in which the article is published. To identify journal articles on a particular topic, you need to use a database. Searching for an article is easier if you already know a bit about the topic, for example, from reading the course textbook, lab manual, and lecture notes. This means you will already know some of the jargon and key terms for the topic before you start. In this case, you can search for articles using key words and refine your search by changing or combining key words, limiting databases or time of publication.

When a topic is mentioned in your textbook or lectures, you can search for a cited reference to find out more details about the topic. Such references in turn might lead you to other cited sources. When reading for your essay/report, you may also find one person's research mentioned frequently in the textbook, and you can then search for works and publications of this person as an author.

REFERENCE GUIDELINE 1:

Select the most relevant references.

The most relevant references consist of the most significant and the most available references. As discussed above, such references are generally journal articles followed by books. Rather than listing any and all papers published on the topic, select the most relevant references by citing original or review articles and choosing the most important paper on a subject (this will also keep the number of your references low.) One indication of the importance of a paper is the number of its citations compared to other articles on the topic. The number of citations is usually provided information in the results of a literature search, and—for quick reference—can also be gleaned when doing a Google search. To validate specific findings, use a primary source, which is the original, peer-reviewed publication of a scientist's new data, results, and theories. For general overview of a topic, you may also use secondary sources (e.g., a review article) or certain tertiary sources such as textbooks.

REFERENCE GUIDELINE 2:

Verify your references against the original document.

References tend to have a surprisingly high rate of error. Therefore, when you use references found in other sources, you need to verify them against the original document. Make sure you have read all references you cite to prevent false representation of the reference or the information within. In addition, ensure that every reference in the text is included in the Reference List and that every reference in the Reference List is cited in the text. Ensure also that citations and references follow the format requested in any instructions for composing your document.

2.3   CITING REFERENCES

Whenever you use the ideas and findings of others, the source needs to be cited in the text and listed in a Reference List at the end of an article or paper.

REFERENCE GUIDELINE 3:

Know where to place references in a lab report or scientific paper.

Abstract   Avoid placing any citations in the Abstract. Start citing sources in your Introduction.

Introduction   Cite the most relevant references only. Although the amount of background information needed depends on the audience, do not review the literature.

Materials and Methods   When appropriate, cite original references for methods used in your study. (For example: "Growth was measured and analyzed according to Billings (1988).")

Results   Usually statements that need to be referenced are not written in the Results section. Comparison statements are made in the Discussion section.

Discussion   Include references to compare and contrast your findings, studies that provide explanations, or those that give your findings some importance.

REFERENCE GUIDELINE 4:

Cite references in the requested form and order.

References are listed in two formats in scientific documents: as text citations and in the Reference List. Text citations list references within the text in short version, such as by name and year. The Reference List at the end of a document displays the full citation of your reference.

In the text as well as in the Reference List, references can be cited in different ways. Some common formats for text citations are "(author, year)," "(number)," and "number". Always check the guidelines for any document you have to complete to ensure that your citations are correctly formatted. Two examples of text citations are shown in Example 2–1a and 2–1b.

Example 2–1 a   Vit-E is a fat-soluble vitamin (**Hollander et al., 1976**).

b   Vit-E is a fat-soluble vitamin (**8**).

If you cite multiple references for a point in your text, list the references in chronological order.

Generally, for a publication by one author, cite that author's name:

Example 2–2a   . . . described by Popi (18, 20).

For a paper by two authors, cite both authors' names:

Example 2–2b   Daniles and Ebert (9) reported XYZ.

For a paper by three or more authors, cite the first author's name followed by "et al."

Example 2–2c   . . . has previously been reported (Brown et al., 1999a; Brown et al., 1999b; Liu et al., 2003).

REFERENCE GUIDELINE 5:

Know where to place references in a sentence.

References can be incorporated into the text in various ways. In general, references can be placed either after the idea you are referring to or after the names of the authors depending on what you want to emphasize: the science or the scientist.

Example 2–3 a   Starfish fertilization is species specific (**17**).

b   Peterson (**17**) reported that starfish fertilization is species specific.

Do not place references in the middle of an idea or after general information of a study, such as after "in a recent study" or "has been reported."

Also note that references for different points in one sentence have to be cited after the appropriate point rather than grouping all the references together at the end of the sentence.

Example 2–4

Compound A can be separated from the mixture by two methods: distillation **(Ramos et al., 2011)** and HPLC **(Koehler et al., 2004)**.

REFERENCE GUIDELINE 6:

List references in the requested style in the reference list.

Your Reference List at the end of your paper (commonly also referred to as "Literature cited") should contain a list of only the literature cited in the text. Be sure to check and follow the require reference style carefully. Reference styles vary on such details as to whether titles of articles are included, whether last page numbers are included, where authors' initials are placed (before or after the last name), where the year of publication is placed (after the authors' names, after the journal title, at the end of the reference), and how items are punctuated.

Typically, if you used the "(author, year)" system in the text, references are listed in alphabetical order and are not numbered in the Reference List.

Example 2–5a

Bailey, S.E., Olin, T.J., Bricka, R.M., and Adrian, D.D. (1999). A review of potentially low-cost sorbents for heavy metals. *Water Res.* **33:**2469–2479.

Das, N.C. and Bandyopadhyay, M. (1992). Removal of copper(II) using vermiculite. *Water Environ. Res.* **64:**852–857.

Hani, H. (1990). The analysis of inorganic pollutants in soil with special regard to their bioavailability. *J. Environ. Anal. Chem.* **39:**197–208.

Lackovic, K., Angove, M.J., Wells, J.D., and Johnson, B.B. (2004). Modelling the adsorption of Cd(II) onto goethite in the presence of citric acid. *J. Colloid Interface Sci.* **269:**37–45.

If you used the (number) system in the text, references in the list are numbered in the order in which each reference is first cited in the text.

Example 2–5b

1. Lackovic, K., Angove, M.J. Wells, J.D. and Johnson, B.B. (2004). Modelling the adsorption of Cd(II) onto goethite in the presence of citric acid. *J. Colloid Interface Sci.* **269**:37–45.

2. Bailey, S.E., Olin, T.J., Bricka, R.M., and Adrian, D.D. (1999). A review of potentially low-cost sorbents for heavy metals. *Water Res.* **33**:2469–2479.

3. Hani, H. (1990). The analysis of inorganic pollutants in soil with special regard to their bioavailability. *J. Environ. Anal. Chem.* **39**:197–208.

4. Das, N.C. and Bandyopadhyay, M. (1992). Removal of copper(II) using vermiculite. *Water Environ. Res.* **64**:852–857.

References to books or book chapters also require great attention to detail. Here too, there are various formats and you may have to adjust yours to the one requested in class.

Example 2–6

Ege, Seyhan N. (1984). *Organic Chemistry.* D.C. Heath and Company, Lexington, Massachusetts/ Toronto, pp. 203–229.

2.4 COMMON REFERENCE STYLES

If no citation instructions are given, consider using one of the more commonly accepted formats, such as the Modern Language Association (MLA), Council of Scientific Editors (CSE; formerly CBE), or American Psychological Association (APA) style. Another option is to select a pertinent journal in that field and then use the format of that journal in your write-up.

Example 2–7

MLA Style

In Text . . . as reported previously (1).

Bibliography **Books**
Okuda, Michael, and Denise Okuda. *Star Trek Chronology: The History of the Future.* New York: Pocket, 1993.

Journal Article
Jefferson, Thomas A., Stacey, Pam J., and Baird, Robin W. "A review of Killer Whale interactions with other marine mammals: predation to co-existence." *Mammal Review* 21.4 (2008):151–80.

Example 2–8	CSE Style	
	In Text	. . . observed by Hinter (2008), (McCormac and Kennedy 2004) (Meise et al. 2003) (Hinter 2008)
	Bibliography	**Books** McCormac JS, Kennedy G. 2004. *Birds of Ohio.* Auburn (WA): Lone Pine. p. 77–78.
		Journal Article Meise CJ, Johnson DL, Stehlik LL, Manderson J, Shaheen P. 2003. Growth rates of juvenile Winter Flounder under varying environmental conditions. *Trans Am Fish Soc* 132(2):225–345.

Example 2–9	APA Style	
	In Text	. . . as reported by Juls (1999). Research by Wegener and Petty (1994) supports . . . (Juls, 1999) (Wegener & Petty, 1994) (Kernis et al., 1993)
	Bibliography	**Books** Calfee, R. C., & Valencia, R. R. (1991). *APA guide to preparing manuscripts for journal publication.* Washington, DC: American Psychological Association.
		Journal Article Scruton, R. (1996). The eclipse of listening. *The New Criterion, 15*(30), 5–13.

REFERENCE GUIDELINE 7:

Know how to cite and list references from the Internet.

Use of Web citations is not always accepted, but this is a developing area. Generally, you as the author can decide which style to choose. However, make sure that the style does not conflict with that asked for by your instructor.

To cite and list a reference from the Internet or the World Wide Web, use the following form:

Example 2–10	Author's name (last name first). Title. Available from: URL: http://Internet address or World Wide Web address.

There are also other good reference styles for citing Internet sources such as the MLA, CSE (formerly CBE), or Chicago styles.

Example 2–11

MLA Style

Online document
Author's name (last name first). Document title. Date of Internet publication. Date of access <URL>.

Book
Bryant, Peter J. "The Age of Mammals." *Biodiversity and Conservation.* 28 Aug. 1999. 4 Oct. 1999 <http://darwin.bio.uci.edu/~sustain/bio65/lec02/b65lec02.htm>.

Article in an electronic journal (ejournal)
Joyce, Michael. "On the Birthday of the Stranger (in Memory of John Hawkes)." *Evergreen Review* 5 Mar. 1999. 12 May 1999 <http://www.evergreenreview.com/102/evexcite/joyce/nojoyce.html>.

Example 2–12

CSE Style

To document a file available for viewing and downloading via the **World Wide Web**, provide the following information:
- Author's name (if known)
- Date of publication or last revision
- Title of document
- Title of complete work (if relevant)
- URL, in angle brackets
- Date of access

Book
Bryant P. 1999 Aug 28. Biodiversity and conservation. <http://darwin.bio.uci.edu/~sustain/bio65/index.html>. Accessed 1999 Oct 4.

Article in an electronic journal (ejournal))
Browning T. 1997. Embedded visuals: student design in Web spaces. Kairos: A Journal for Teachers of Writing in Webbed Environments 3(1). <http://english.ttu.edu/kairos/2.1/features/ browning/bridge.html>. Accessed 1997 Oct 21.

Example 2–13

Chicago Style

To document a file available for viewing and downloading via the **World Wide Web**, provide the following information:
- Author's name
- Title of document, in quotation marks
- Title of complete work (if relevant), in italics or underlined
- Date of publication or last revision
- URL, in angle brackets
- Date of access, in parentheses

Book
Peter J. Bryant, "The Age of Mammals," in *Biodiversity and Conservation* August 1999. <http://darwin.bio.uci.edu/~sustain/bio65/index.html> (October 4, 1999).

Article in an electronic journal (ejournal)
Tonya Browning, "Embedded Visuals: Student Design in Web Spaces," *Kairos: A Journal for Teachers of Writing in Webbed Environments* 3, no. 1 (1997), <http://english.ttu.edu/kairos/2.1/features/browning/index.html> (October 21, 1999).

2.5   MANAGING SOURCES

REFERENCE GUIDELINE 8:

Manage your references well.

Keep a list of references to help organize and keep track of them. You can save yourself much time and much frustration if you manage your references from the start. If you have many references in your list, such as when you are writing an article for publication, computer programs that put references in various formats are available (EndNote or Reference Manager) and are very useful. In these cases, start using a reference managing program right when you download your references from the library. There are few aspects of preparing a manuscript that are more irritating than painstakingly typing, changing, or correcting the Reference List. If you are unfamiliar with such a program, inquire at your library. Most libraries offer short classes on reference programs.

2.6   PLAGIARISM AND TAKING NOTES

REFERENCE GUIDELINE 9:

Ensure that you are not plagiarizing.

In scientific writing, direct quotations are rarely used. Instead, information is commonly summarized and paraphrased. In all cases, the source has to be cited. Failing to indicate the source of information in scholarly scientific work is called plagiarism and is a form of academic misconduct. You are obligated, as an ethical responsibility to other writers and as a defense for yourself, to acknowledge all borrowings you take from other sources, even if you do not copy the exact words used in the original. This rule is usually also pointed out in a school's honor code of conduct and covered during orientation.

To avoid plagiarism, you need to know what constitutes it. Plagiarism includes:

- Quoting material without acknowledging the source. (This is the most obvious kind of plagiarism.)
- Borrowing someone else's ideas, concepts, results, and conclusions and passing them off as your own without acknowledging them—even if these ideas have been substantially reworded.

- Summarizing and paraphrasing another's work without acknowledging the source.

The rules apply to both textual and visual information. If you are using the Internet as a source of information, you must also cite that source (see Section 2.4).

Know that you do not have to document facts that are considered common knowledge. Common knowledge is information that can be found in numerous places and is likely to be known by a lot of people such as the information found in Example 2–14.

Example 2–14   Many endemic species exist on the Galapagos Islands.

However, information that is not generally known and ideas that interpret facts have to be referenced as in Example 2–15.

Example 2–15   Based on a recent study, the blue iguanas of the Grand Cayman Islands are an endangered species (9).

The finding that "blue iguanas of the Grand Cayman Islands are an endangered species" is not a fact but an *interpretation*. Consequently, you need to cite your source. Similarly, if you are uncertain whether something falls into the common knowledge category, it is best to document it.

Following are some other examples of common knowledge that do not require a citation:

Example 2–16 a   Because nitrogen (N) is an essential component of proteins, enzymes, and other biologically important compounds, the natural cycling of N in freshwater ecosystems needs to be studied in detail.

b   Volcanic eruptions are often preceded and accompanied by "volcanic unrest," providing early warning of a possible impending eruption.

Statements that contain information and interpretations that need to be cited are shown in the next examples.

Example 2–16 a   While low nitrogen (N) availability can limit primary productivity, excess N has been linked to eutrophication and health concerns (19).

b   The eruption of Mount Pinatubo in 1991 was preceded by a relatively short progression of precursory activity before its full-blown eruption (22).

Know that imitation and borrowing by themselves are not plagiarism. Drawing on other people's ideas is perfectly reasonable and in fact unavoidable when you write academic documents—but you must acknowledge the source.

REFERENCE GUIDELINE 10:
Keep track of ideas and references.

REFERENCE GUIDELINE 11:
Know how to paraphrase.

Paraphrasing means taking another person's ideas and putting those ideas in your own words. This is the skill you will probably use most when incorporating sources into your writing. Although you use your own words to paraphrase, you must still cite the source of the information in the text at the end of the sentence or idea.

It is important that you distinguish between paraphrasing and plagiarizing. Changing a word or two in someone else's sentence or changing the sentence structure while using the original words is not paraphrasing but plagiarizing. The best way to avoid plagiarism is to do the following when collecting and using information in scientific writing:

- Keep track of references and write down the information you intend to use whenever you come across a passage that you think may be useful for your document.
- Keep a detailed list of sources.
- If you copy something word by word, put it in quotation marks, but know that writing in the sciences uses direct quotations only rarely. When you want to use details from the original but not necessarily all of them and not necessarily in the same order as the original, you need to paraphrase.
- Write down the most important ideas in your own words using bullet points.
- Take notes with the book closed. This way you are forced to put the ideas into your own words.
- Double-check that the reference and information is correct by going back to the original when you compose your document.

Following is an example of a paragraph that instead of being paraphrased has been plagiarized:

Example 2–17   **Plagiarized paragraph**

Original:
Healthy older adults often experience mild decline in some areas of cognition. The most prominent cognitive deficits of normal aging include forgetfulness, vulnerability to distraction and other types of interference, as

well as impairments in multi-tasking and mental flexi-
bility. These cognitive functions are the domain of the
prefrontal cortex, the most highly evolved part of the
human brain. Prefrontal cortical cognitive abilities begin
to weaken even in middle age, and are especially im-
paired when we are stressed. Understanding how the
prefrontal cortex changes with age is a top priority for
rescuing the memory and attention functions we need
to survive in our fast-paced, complex culture.

Plagiarized sample:
In healthy older adults, often some areas of cognition decline. The most noticeable cognitive declines
of normal aging include forgetfulness, vulnerability
to distraction, and problems in multi-tasking. These
cognitive tasks are localized in the prefrontal cortex,
which is the most highly evolved portion of the brain.
Already in middle age, prefrontal cortical cognitive
functions start to decrease. Such functions are also
particularly affected during any type of stress. Studying memory and attention is important to understand
how the prefrontal cortex changes with age. It is particularly important to understand these changes in our
current fast-paced lifestyle.

The preceding passage is considered plagiarism for two reasons: (a) the
writer has only changed around a few words and phrases or changed the
order of the original's sentences, and (b) the writer has failed to cite a source
for any of the ideas or facts.

An acceptable way of paraphrasing the preceding sample paragraph
would be the following:

**Revised
Example 2–17**

Paraphrased sample:
A recent study shows that the process of aging is accompanied by a decline in cognitive abilities, deficits
in working memory, and compromised integrity of
neural circuitry in the brain (3). Impairments in these
executive operations compromise the regulation of
thought, emotion, and behavior, and ultimately jeop-
ardize independence and quality of life. Elucidating
mechanisms underlying the decline of neural circuitry
is critical for the treatment of debilitating cognitive
deficits, such as those found in the elderly.

This is acceptable paraphrasing because the writer accurately relays the
information in the original using his or her own words. The writer also lets
the reader know the source of the information.

Unlike elsewhere in a scientific research paper, many portions of the
Materials and Methods section will sound extremely similar to each other,
mainly because there are only so many ways one can describe procedures

whose technique and setup is essentially identical with the exception of the variables. Using very similar phrases in such passages, along with substituting your variables, would not be considered plagiarism. Therefore, do not desperately try to invent new wordings to describe the same procedure.

Here are some examples of passages that would not be considered plagiarized:

Example 2–18

Method description in paper A:
Real-time fluorescence quantitative PCR was performed in an Applied Biosystems Prism 7000 instrument in the reactions containing an Applied Biosystems SYBR green master mix reagent and oligonucleotide pairs to the endogenous control gene 'A' and cDNA of 'B'. The reagents were denatured at 95°C for 10 min, followed by 40 cycles of 15 s at 95°C and 60 s at 60°C. The primer sequences (5'–3') were 'A' forward 5'-GACACCTATGCCGAACCGTGAA-3'; 'A' reverse, 5'-CTGAGTATCAGTCGGCCTTGAA-3'; 'B' forward 5'-GTTCGACGACATCAACATCA-3'; 'B' reverse 5'-TGATGACGTCCTTCTCCATG-3'.

Method description in paper B:
PCR amplification of 'X' sequences was done using the GC RICH PCR System (Roche, Mannheim, Germany). All non-'X' sequences were amplified using Taq DNA polymerase (Promega, Madison, WI). Primers were designed using published sequences for 'X-1' (GenBank: Xxxxxx) and 'x-13' (GenBank: Xxxxxx) (Table 1). PCR thermal cycling conditions were: 2 min at 50°C, 10 min at 95°C, followed by 40 cycles of 15 s at 95°C and 1 min at 60°C. PCR reactions were run with molecular weight standards on 0.8% agarose gels containing ethidium bromide and visualized by UV light. The primers used were: 5'-GGCTCAC-CAGCATCATATACG-3' and 5'-GGCTACAATGAC-GACGTCA-3'.

Method description in paper C:
Real-time PCR was performed using the TaKaRa SYBR PCR kit and ABI Prism 7000 sequence detection system according to the manufacturer's specifications. The primers for amplification were *abc* (5'-CGCTCCTCTGCATCTAAT-CAG-3' and 5'-GACACTTAGCACGCACTCA-3') and *def* (5'-GCATCTTCAAGTAAGGACTATC-3' and 5'-GACTTTCACAGTACCAGATT-3'). Total reaction volume was 50 μl including 25 μl SYBR Premix Ex Taq with SYBR Green I, 300 nM forward and reverse primers and 2 μl cDNA. The thermal cycler program was 1 cycle at 95°C for 10 s, followed by 40 cycles at 95°C for 5 s and 60°C for 30 s. The PCR products were detected by electrophoresis through a 2% agarose gel stained with ethidium bromide.

For these passages and for similar ones that occur mainly in the Materials and Methods section of a research paper, it may not be a bad idea to collect sample phrases from other articles for your reference. Know, however, that I am not advising you to copy entire passages to be placed into your manuscript but instead individual sample phrases and expressions that can be applied to writing your research article.

SUMMARY

REFERENCE GUIDELINES:
1. Select the most relevant references.
2. Verify your references against the original document.
3. Know where to place references in a scientific paper.
4. Cite references in the requested form and order.
5. Know where to place references in a sentence.
6. List references in the requested style in the reference list.
7. Know how to cite and list references from the internet.
8. Manage your references well.
9. Ensure that you are not plagiarizing.
10. Keep track of ideas and references.
11. Know how to paraphrase.

PROBLEMS

Problem 2-1

In the following paragraph, which is part of an Introduction, check that the text citations have been placed appropriately.

Ostracodes are small bivalved Crustacea that form an important component of deep-sea meiobenthic communities along with nematodes and copepods (10). Crustaceans (10, 11) are dense and diverse in the deep sea and are one of the most representative groups of whole deep-sea benthic community. Pedersen et al. as well as Jackson et al. reported (12–14) that ostracode species have a variety of habitat and ecology preferences (e.g., infaunal, epifaunal, scavenging, and detrital feeders), representing a wide range of deep-sea soft sediment niches. Furthermore, Ostracoda is the only commonly fossilized metazoan group in deep-sea sediments. Thus, fossil ostracodes are considered to be generally representative of the broader benthic community. The distribution and abundance of deep-sea ostracode taxa in the North Atlantic Ocean are influenced by several factors (14, 15), among them, temperature, oxygen, sediment flux, and food supply. Several paleoecological studies suggest (1, 16) that these factors influence deep-sea ecosystems over orbital and millennial timescales.

(With permission from National Academy of Sciences, U.S.A.)

Problem 2-2

Paraphrase the following passage.

A drought is defined in the glossary of meteorology as "a period of abnormally dry weather sufficiently prolonged so that the lack of water causes a serious hydrologic imbalance in the affected area." A drought is the consequence of a natural reduction in the amount of precipitation received over an extended period of time, usually a season or more in length, although the onset and end of a drought are generally difficult to determine. Drought is a normal, recurring feature of climate, which occurs in high, as well as low rainfall areas. Although it is a frequent and often catastrophic feature in semiarid climates, it is less frequent and disruptive in humid regions and an even less meaningful concept for deserts. The severity of a drought can be significantly aggravated by other climatic factors, such as by high temperatures and winds or by low relative humidity. Resulting imbalances may include crop damage, water supply shortage, and increased fire hazards. These effects often accumulate slowly over a considerable period of time and may linger for years after drought termination. There remains much disagreement within the scientific and policy community about exact drought characteristics, and therefore not much progress has been made in drought management in many parts of the world.

(Miller. J. and Carling, C. 2012. Journal X, pp. 54–55.)

Problem 2-3

Paraphrase the following passage.

A few simple mutations of the deadly H5N1 avian influenza virus might give it the ability to spread readily among humans. Once in the human population, the virus could spark a global pandemic that could kill tens of millions. The virus started spreading in earnest among birds in late 2003. It spread to various countries and appears to be here to stay. In 2007, the virus surfaced in poultry flocks in eight new countries, including in Bangladesh, Poland, and Ghana. Outbreaks returned in 23 countries stretching from Indonesia to the United Kingdom. In some countries, such as in Indonesia and Nigeria, outbreaks are now more or less continuous. Indonesia is the hardest-hit country, reporting 42 cases and 32 deaths in 2007. As long as the virus is circulating in birds, sporadic human cases will continue to occur, and most of them will be fatal. Although poultry trading is the primary means of spreading the virus, the role wild birds play in long-distance spread is still unclear. Therefore, scientists should take advantage of new lightweight transmitters that enable satellite tracking of migratory species.

(Smith. J. (2012). Journal ABC, vol 34.)

Problem 2-4

The following passage comes from a research paper publication. Read through the original to get an understanding of its central points and then determine whether the student revisions are plagiarized or paraphrased versions.

Original

Hammerhead sharks are well known as predators of octopus, squid, crustaceans, and fish. Attacks have been observed on 32 species of cephalopods, 12 species of crustaceans, 26 species of stingrays, other hammerhead sharks, and even their own young. Various marine mammals, except dolphins and whales, have also been recorded as prey of hammerhead sharks. Ecological interactions have not been systematically studied, and further work may show that the hammerhead shark is a more important predator for some populations than for others. Not all behavioral interactions between hammerhead sharks and marine species result in predation, however. Some involve 'harassment' by the hammerhead sharks, feeding by multiple marine species in the same area, other predators being tolerated around hammerhead sharks, all apparently "ignoring" each other. Some reports also describe attacks of hammerheads by porpoises and killer whales. These nonpredatory interactions are relatively common. Our findings show that interactions between hammerhead sharks and other marine animals are complex, involving many different factors that we are just beginning to understand.

(Source: Haskins, T. A., Nepomucino, P. and Bert, W. (2011). ABX Review vol 26(3))

Student Version A

Hammerhead sharks prey on other marine animals such as on squid and lobster, and on many members of the stingray family (Haskins et al., 2011). Hammerheads have also attacked members of their own family. Some interactions between hammerhead sharks and other marine species do not result in predation, for example, "harassment" by the hammerheads, feeding by different species in the same area, toleration of other species of sharks, and even attacks on hammerheads by dolphins. These nonpredatory interactions are relatively common. Thus, interactions between hammerhead sharks and marine animals are complex, and involve many different factors that we are just beginning to understand. Further work is necessary to understand the ecological interactions of hammerhead sharks.

Student Version B

Ecological interaction of hammerhead sharks with other marine animals can be predatory or nonpredatory. Predatory interactions involve hammerhead attacks on many marine animals. Nonpredatory interactions involve hammerheads harassing other marine animals, feeding in the same area, and even attacks on hammerheads by the other species (Haskins et al., 2011). Although these interactions have been observed, further work is needed to fully understand hammerhead shark interactions with other marine animals.

CHAPTER 3

Fundamentals of Scientific Writing, Part I—Style

Various "rules" of scientific writing have been published and discussed by others. Some concentrate on basic style and composition, others on the psychology of the reader, and yet others on common grammar problems editors encounter. The sources on which many of the basic scientific writing principles discussed in this chapter are based include those of Michael Alley (1996 and 2000), Robert Day and Barbara Gastel (2006), George Gopen (1990), George Gopen and Judith Swan (1990), Diana Hacker (2004), E.J. Huth (1990), Michael Katz (2006), Karin Knisely 2009), Janice Matthews et al. (1996), Victoria McMillan (1997); Maeve O'Connor (1991), Ian Pechenik (2009), Ann Penrose and Stephen Katz (2004), Leslie Perelman et al. (1997), Philip Rubens (1992), W. Stunk and E.B. White (1979); Robert Sternberg (2003), Susan Thurman (2002), Joseph Williams (1988 and 2006), Otto Yang (1995), Petey Young (2006), and Mimi Zeiger (2000) (*see also Bibliography*). This chapter compiles guidelines recommended in their works and those based on my own experience working with students into a comprehensive list of essential scientific writing principles.

3.1   WORD CHOICE AND LOCATION

WRITING PRINCIPLE 1:
Write with the reader in mind.

In the professional world, success in writing is determined by whether your readers understand what you are trying to say. You need to write clearly so that readers can follow your thinking and so that you achieve the highest possible impact. To "write with the reader in mind" means to consider how the reader interprets what you have written. It requires you to construct your writing clearly, concisely, and at the right level, such that the reader can follow and understand what you want to say immediately. This principle should be viewed as the Central Principle around which all other principles revolve.

WRITING PRINCIPLE 2:
Use precise words.

Words in science should be precise. Imprecise word choices can be problematic as they will be unclear to readers. Such word choices should be revised as shown in the following example:

👎	**Example 3-1**	Plants were kept <u>in the cold</u> overnight.

👍	**Revised Example 3-1**	Plants were kept **at 0°C** overnight.

👎	**Example 3-2**	Reagent Y was mixed <u>with</u> X.

👍	**Revised Example 3-2**	Reagent Y and X were mixed together. **OR** Reagent Y was mixed **using** X.

WRITING PRINCIPLE 3:
Use simple words.

Scientific writing has many heavy, fancy technical words. To keep your writing from being too heavy, choose simple words for the rest of the sentence to ensure that as many readers as possible can understand what has been written.

The following revised sentences are much easier understood by readers because their word choice is much simpler.

	Example 3-3	We <u>utilized</u> UV light to induce mutations in *Arabidopsis*.

	Revised Example 3-3	We **used** UV light to induce mutations in *Arabidopsis*.

	Example 3-4	<u>For the purpose of</u> examining cell migration, we dissected mouse brains.

	Revised Example 3-4	**To** examine cell migration, we dissected mouse brains.

WRITING PRINCIPLE 4:

Omit unnecessary words and phrases.

Your writing should be as brief as possible but clear. Many sentences in science appear complex because they contain unnecessary words and phrases or redundancies.

Unnecessary words

	Example 3-5	The sample size was not <u>quite sufficiently large</u> enough.

	Revised Example 3-5	The sample size was not **large** enough. **OR** The sample size was too small.

The following words usually can and should be omitted entirely.

	actually	basically	essentially	fairly	much	really
	practically	quite	rather	several	very	virtually

In the following examples of redundancies, all the words in parentheses can be omitted.

	(already) existing	at (the) present (time)
	blue (in color)	cold (temperature)
	(completely) eliminate	(currently) underway
	each (individual)	each and every (choose one)
	(end) result	estimated (roughly) at
	(final) outcome	first (and foremost)
	(future) plans	never (before)
	period (of time)	(true) facts

Unnecessary phrases

Aside from avoiding unnecessary words, omit unnecessary phrases to make your writing shorter and clearer.

👎	**Example 3-6**	It is well known that there are three subtypes of the KL-2 virus . . .

👍	**Revised Example 3-6**	There are three subtypes of the KL-2 virus . . .

👎	**Example 3-7**	In a previous study it was demonstrated that Iba 1 was detected in monocytic cells.

👍	**Revised Example 3-7**	Iba 1 was detected in monocytic cells (Lopez et al., 1995).

Commonly used unnecessary phrases that can usually be deleted include:

👎
there are many papers stating . . .	it is speculated that . . .
it was shown to . . .	it has been found that . . .
it has long been known that . . .	it has been reported that . . .

Phrases to avoid

Avoid	Better	Avoid	Better
a considerable number of	many	in many cases	often
adjacent to	near	in order to	to
an adequate amount of	enough	in some cases	sometimes
a consequence of	because	in the absence of	without
at a rapid rate	rapidly	in the event that	if
at no time	never	it is clear that	clearly
based on the fact that	because	it is of interest to note that	note that
despite the fact that	although	it is often the case that	often
due to the fact that	due to	majority of	most
during the course of	during, while	no later than	by
during the time that	while, when	on the basis of	by
elucidate	explain	prior to	before
employ	use	referred to as	called
facilitate	enable	regardless of the fact that	even though
first of all	first	so as to	to
for the purpose of	to	terminate	end
has the capability of	can, is able	the great majority of	most
in case	if		

WRITING PRINCIPLE 5:

Avoid too many abbreviations.

Too many abbreviations can be confusing to the reader. Thus, abbreviations should be kept to a minimum. Similarly, nonstandard abbreviations need to be limited; otherwise, the reader will get lost. Nonstandard abbreviations also need to be defined.

| Example 3-8 | The increase in <u>SSTIs</u> caused by <u>CA-MRSA</u> but not by <u>MSSA</u> is especially marked in the pediatric population as a study from <u>PCH</u> shows. |

The previous example may be perfectly intelligible to expert colleagues but will be unintelligible to others.

| Revised Example 3-8A | The increase in **soft tissue infections** caused by **community-acquired methicillin-resistant** *Staphylococcus aureus* but not by **methicillin-sensitive** *S. aureus* is especially marked in the pediatric population as a study from **Philadelphia Children's Hospital** shows. |

If lengthy terms are used often throughout a scientific document, then abbreviations that have been introduced can be used instead. In these cases, define essential nonstandard abbreviations at their first appearance, although not in a title or abstract.

| Example 3-9 | Over the past two decades, **methicillin-resistant** *Staphylococcus aureus* **(MRSA)** has frequently been isolated from hospitalized patients, but also in the community. |

WRITING PRINCIPLE 6:

Use correct nomenclature and terminology.

General Scientific Nomenclature

Use correct vocabulary, nomenclature, and terminology to avoid being misunderstood or confusing the reader. If you are not sure, do not guess. Take the time to look up terms in a dictionary or other reference book. Dictionaries for the biological, medical, and other scientific fields as well as online dictionaries are listed in the back of this book.

The most commonly used scientific nomenclature includes:

Species and most Latin derivates are in *italics (in vivo, Physcomitrella patens)*
Human genes: all caps and *italics (ADH3, HBA1)*

Mouse genes: first letter capitalized, the rest lower case, *italics (Sta, Shh, Glra1)*

Human proteins: capitals, no italics (ADH3, HBA1)

Mouse proteins: like genes, but no italics (Sta, Shh, Glra1)

To distinguish the species of origin for homologous genes with the same gene symbol, an abbreviation of the species name is added as a prefix to the gene symbol—for example, human loci: (HSA)*G6PD*; homologous mouse loci: (MMU)*G6pd,* where HSA = *Homo sapiens,* MMU = *Mus musculus.*

Restriction enzymes: a combination of *italics* and nonitalics (e.g., *Bam* HI). Check supplier.

Misused and Confused Terms

Words are not always what they seem. Quite a few words and expressions in science are commonly misused and confused. Watch out for these misused and confused scientific terms. Consult a dictionary when you write if you are unsure. The following brief table lists the most commonly confused terms. A more comprehensive list can be found in Appendix A.

affect	verb, meaning "to act on" or "to influence" (note: affect is often also used as a noun, meaning emotion or desire, in psychology)	The addition of KI-3 to MZ1 cells **affected** their growth rate (i.e., it could have increased or decreased or induced).
effect	1. noun, meaning a result or resultant condition 2. verb, meaning "to cause" or "to bring about"	1. We examined the **effect** of KI-3 on MZ1 cells. 2. The addition of KI-3 to MZ1 cells **effected** a chance of growth rate (i.e., it could be caused or brought about).
as	conjunction, used before phrases and clauses	Let me speak to you **as** a father (= I am your father and I am speaking to you in that character)
		as we just mentioned **as** described previously, **as** described by…
like	preposition, meaning "in the same way as"	Let me speak to you like a father (= I am not your father but I am speaking to you as your father might). Our observations were **like** those of Andrews et al.
which	use with commas for nondefining (nonessential) sentences	Dogs, which are cute, recovered.
that	use without commas for essential sentences. A phrase or clause introduced by *that* cannot be omitted without changing the meaning of the sentence.	Dogs that were treated with antidote recovered.

contrary to	preposition meaning "in opposition to"	**Contrary to** our expectations, addition of Mg^{++} did not alter our results.
on the contrary	usually used only in spoken English; used when one says a statement is not true	"It's exciting!" "**On the contrary**, it's boring!"
on the other hand	rarely used in scientific English; used when adding a new and different fact to a statement	On the one hand, plants need sunshine, and **on the other hand** they need rain.
in contrast	used for two different facts that are both true, but pointing out the surprising difference between them	pH values increased for prokaryotes. **In contrast**, no pH difference was observed in eukaryotes

WRITING PRINCIPLE 7:

Establish importance.

Word location within a sentence can help authors to guide and influence readers. The format and structure authors use to present information will lead the reader to interpret it as important or less important. To decide on the best placement of words within a sentence, it is crucial that authors establish importance. In general, the end position in a sentence is more emphasized than the beginning position, and the main clause is more emphasized than the dependent clause.

Consider the following two versions of a sentence:

Example 3–10	a	Although vitamin B6 reduces the risk of macular degeneration, taking it has some side effects.
	b	Although taking vitamin B6 has some side effects, it reduces macular degeneration.

Because of the placement of information in these sentences, most readers view the overall recommendation of vitamin B6 as negative in sentence a, whereas that of sentence b is viewed by most as positive.

WRITING PRINCIPLE 8:

Place old, familiar, and short information at the beginning of a sentence.

WRITING PRINCIPLE 9:

Place new, complex, or long information at the end of a sentence.

Readers expect to see old information that links backward at the beginning of a sentence (or paragraph) and new information at the end of a sentence (or paragraph) where it is more emphasized. If information is placed where most readers expect to find it, it is interpreted more easily and more uniformly.

If, in addition, the information is linked through word location, such as when the information at the end position of a sentence is placed at the beginning, or topic position, of the next sentence, writing "flows" much better, as in the following example.

Example 3-11

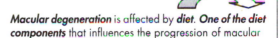

Macular degeneration is affected by *diet*. *One of the diet components* that influences the progression of macular

degeneration is *vitamin B6*. Although *vitamin B6* seems to reduce the risk of macular degeneration, it may have some *side effects*.

WRITING PRINCIPLE 10:
Get to the subject of the main sentence quickly.

Readers understand a sentence more easily if its subject is readily available. When a sentence is started with several words before its subject/topic, readers find it hard to see what the sentence is about. Avoid such long introductory phrases and subjects.

Example 3-12   **Because high-resolution EM methods can show fine details of oligomeric proteins or complexes such as ribosomes**, *these methods* may show where nucleic acid strands are located.

Revised Example 3-12   *High-resolution EM methods* may show where nucleic acid strands are located because these methods can show fine details of oligomeric proteins or complexes such as ribosomes

---

WRITING PRINCIPLE 11:
Place the verb immediately after the subject and the object right after the verb.

Readers expect grammatical subjects to be followed immediately by the verb. Anything of length that intervenes between subject and verb is read as an interruption, and therefore as something of lesser importance.

Example 3-13	<u>Dengue virus</u>, a Flavivirus belonging to the family *Flaviviridae*, which includes over 60 known human pathogens such as those causing yellow fever, Japanese encephalitis, tick-borne encephalitis, Saint Louis encephalitis, and West Nile encephalitis, <u>is classified</u> into four different serotypes, types 1, 2, 3, and 4.

This sentence obstructs the reader, because the grammatical subject ("Dengue virus") is separated from its verb ("is classified") by 31 words.

Often an interruption can be moved to the beginning or the end of a sentence, depending on whether it is connected to old or new information in the sentence. At other times, the author should consider splitting the information into two sentences or omitting it altogether if it is not essential.

Revised Example 3-13A	***Dengue virus is*** a *Flavivirus* belonging to the family *Flaviviridae*, which includes over 60 known human pathogens such as those causing yellow fever, Japanese encephalitis, tick-borne encephalitis, Saint Louis encephalitis, and West Nile encephalitis. ***It is classified*** into four different serotypes, types 1, 2, 3, and 4.

Revised Example 3-13B	***Dengue virus is classified*** into four different serotypes, types 1, 2, 3, and 4.

Readers also like to get past the verb to the object of a sentence quickly. Thus, authors should avoid any interruptions between verb and object by placing interrupting passages either at the beginning or at the end of the sentence.

Example 3-14	We have appended, <u>to determine if deoxyribozymes can similarly be activated by effector molecules</u>, an anti-adenosine aptamer to a selected deoxyribozyme ligase.

Revised Example 3-14	**To determine if deoxyribozymes can similarly be activated by effector molecules**, we have appended an anti-adenosine aptamer to a selected deoxyribozyme ligase.

3.2   SENTENCES AND TECHNICAL STYLE

WRITING PRINCIPLE 12:

Use the first person.

It once was fashionable to avoid using "I" or "we" in scientific research papers because these terms are subjective, whereas the aim in science is to be objective. Contemporary scientific English recognizes that science is not purely objective. Thus, use the first person ("I" or "we") for describing what you did.

☜	**Example 3-15**	It is believed that . . .

☝	**Revised Example 3-15**	**We believe** . . .

☜	**Example 3-16**	The authors would like to thank Peter Fefergon.

☝	**Revised Example 3-16**	**We would like** to thank Peter Fefergon.

The use of the first person is more controversial in the Methods section than in the rest of the paper. In this section, materials and methods are usually the topic. In addition, often it may not have been the author(s) who performed a certain experiment but rather a technician or hired helper. Therefore, in the Materials and Methods section, use of third person and passive voice (the form of the verb used when the subject is being acted on rather than doing something) is often preferred.

WRITING PRINCIPLE 13:

Use the active voice.

Distinguish between active and passive voice. When you write in passive voice, the subject is being acted on rather than actively doing something (For example: *Active Voice*: The dog bit the cat. *Passive Voice*: The cat was bitten by the dog.) If the passive voice is used excessively, writing becomes very dull and dense. Therefore, use the active voice rather than the passive voice.

☜	**Example 3-17**	Parrots were attacked by hawk B3.

☝	**Revised Example 3-17**	Hawk B3 attacked parrots.

However, do not remove the passive completely. Use it when readers do not need to know who performed the action or when it sounds more natural, such as in the Materials and Methods section.

WRITING PRINCIPLE 14:

Use past tense for observations, completed actions, and specific conclusions.

Many scientific authors seem to be confused about when to use past tense and present tense. Generally, you should use the past tense for observations, completed actions, and specific conclusions.

   Example 3-18   The IV **caused** local irritation in 53% of the patients.

WRITING PRINCIPLE 15:

Use the present tense for generalizations and statements of general validity.

Use the present tense for generalizations and statements of general validity.

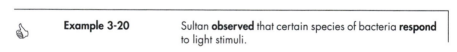   **Example 3-19**   Most regions where this problem <u>arose belonged</u> to
                                  category X.

   Revised   Most regions where this problem **arises belong** to
             Example 3-19   category X.

A sentence can also have mixed tenses, as is apparent in the next example:

   Example 3-20   Sultan **observed** that certain species of bacteria **respond**
                                  to light stimuli.

Here, the experiment has been completed. "Observed" is therefore written in past tense. "Respond", however, is present tense because this part of the sentence is still true and considered established knowledge because it has been published.

WRITING PRINCIPLE 16:

Write short sentences. Aim for one main idea in a sentence.

Short sentences are easier to understand than long sentences. Aim for sentences with an average number of words of about 20–22. Make sure your sentences do not contain more than one main idea. Short, simple sentences tend to emphasis the idea contained in them. However, using only short sentences does not result in strong writing. Some sentences will be long to communicate complex ideas.

Example 3-21
(79 words)

Skin swaps were cultured at the time of the removal of the catheter from seven patients with catheter-related bloodstream infection and in five of these cases (71%) the culture yielded bacteria of the same species with a DNA fingerprint pattern similar to that of the isolates from the catheter and the blood, whereas a different organism grew from skin cultures in the other two patients (29%), suggesting that catheter infection may have originated from contamination of the catheter hub.

Revised
Example 3-21

Skin swaps were cultured at the time of the removal of the catheter from seven patients with catheter-related bloodstream infection. In five of these cases (71%) the culture yielded bacteria of the same species with a DNA fingerprint pattern similar to that of the isolates from the catheter and the blood. However, a different organism grew from skin cultures in the other two patients (29%). These observations suggest that catheter infection may have originated from contamination of the catheter hub.

WRITING PRINCIPLE 17:

Use active verbs.

Verbs are perhaps the most important part of an English sentence. If the action of a sentence is expressed by the main verb, the sentence is natural, direct, and easy to understand. If, instead, the action is expressed in a noun, the verb becomes buried and weak, and the sentence is dense and more difficult to understand. Avoid such nominalizations. Use active verbs instead.

Example 3-22

An <u>increase</u> in temperature <u>occurred</u>.

Revised
Example 3-22

Temperature **increased**.

In the revised sentence, the verb is active and strong. Thus, this sentence is simpler, more direct, and more efficient than when the action is nominalized.

WRITING PRINCIPLE 18:

Avoid noun clusters.

Avoid clusters of nouns that are strung together to form one term. Noun clusters are awkward and often incomprehensible. When nouns appear one right after the other, it can be difficult to tell how they relate to each other and what the real meaning of the cluster is. Instead, use prepositions to link the nouns. Prepositions add clarity to a phrase—they show more fully how the nouns relate to one another—and the meaning of your words becomes clearer.

👎	**Example 3-23**	cultured rat tracheal endothelial cells

👍	**Revised Example 3-23**	cultures **of** endothelial cells **from** the tracheas **of** rats

👎	**Example 3-24**	The strips were exposed to Leishmaniasis disease patients' sera.

👍	**Revised Example 3-24**	The strips were exposed to **sera of patients with Leishmaniasis disease**.

Note that some noun pairs and clusters, such as "water bath" and "sucrose density gradient," are recognized as single words. When you untangle noun clusters, treat such terms as single words.

WRITING PRINCIPLE 19:

Use clear pronouns and prepositions.

Pronouns

Unclear pronouns are one of the most common problems in scientific writing. If the pronoun that refers to a noun is unclear, the reader may have trouble understanding the sentence. Pronouns you use have to refer clearly to a noun in the current or previous sentence. If the pronoun can refer to too many possible nouns, repeat the reference noun after the pronoun.

👎	**Example 3-25**	Gram+ bacteria do not respond to these drugs. Thus, they were of no interest to us.

👍	**Revised Example 3-25A**	Gram+ bacteria do not respond to **these drugs**. Thus, **these drugs** were of no interest to us.

	Revised Example 3-25B	**These drugs** were of no interest to us because Gram⁺ bacteria do not respond to **them**.

Sometimes the noun that a pronoun refers to has been implied but not stated. To clarify the reference, explicitly state the implied noun after the pronoun, as in the next example.

	Example 3-26	If a specimen is frozen in a bath containing dry ice and acetone, the water of the cell can be removed by sublimation and damage to the cell prevented. *This* is commonly used for preservation of cultures.

	Revised Example 3-26	If a specimen is frozen in a bath containing dry ice and acetone, the water of the cell can be removed by sublimation and damage to the cell prevented. ***This technique*** is commonly used for preservation of cultures.

Prepositions

Most verbs can be used with more than one preposition. Be sure to choose the preposition that reflects your intended meaning.

	Example 3–27 a	The human brain is sometimes **compared to** a computer.
	b	When we **compared** our results **with** those of Paulings, et al.....

The most commonly (mis)used prepositions in scientific writing include:

in connection **with**	compared **to/with**
in contrast **to**	correlated **with**
similar **to**	analogous **to**
comparison **of** A **with** B	comparison **between** A **and** B

WRITING PRINCIPLE 20:

Use correct parallel form.

Lists and ideas that are joined by "and," "or," or "but" are of equal importance in a sentence and should be treated equally by writing them in parallel form. To write ideas in parallel form, the same grammatical structures are used.

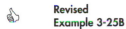

Example 3-28

	subject	verb	prepositional phrase
	The metabolic rate of A	**increased**	**2-fold**
but	**metabolic rate of B**	**decreased**	**10-fold.**

Example 3-29

	preposition	object
The moss *P. patens* grows and	**in** **at**	**the mountains** **sea level.**

Example 3-30   Prolonged fever together with subcutaneous nodules in a child could be due to an infection with a Gram⁺ organism, but it could also be that the child suffers from rheumatic disease.

In this sentence, the groups of words before and after "but" are not parallel, so it is not immediately obvious that the second half of the sentence is giving another possible reason for the illness. The second half of the sentence should be written in parallel form because it is equal in logic and importance.

Revised   Prolonged fever together with subcutaneous nodules
Example 3-30A   in a child **could be due to** an infection with a Gram⁺ organism, but it **could also be due to** rheumatic disease.

This sentence can be further simplified:

Revised   Prolonged fever together with subcutaneous nodules in
Example 3-30B   a child could be **due to** an infection with a Gram⁺ organism or **due to** rheumatic disease.

WRITING PRINCIPLE 21:
Avoid faulty comparisons.

Faulty comparisons and incomplete comparisons are one of the most common problems in scientific writing, as is the overuse of "compared to."

Faulty comparisons arise due to ambiguous comparisons that "compare apples and oranges" instead of like items.

Example 3-31   These <u>results</u> are similar to <u>previous studies</u>.

Revised   These **results** are similar to **the results of** previous
Example 3-31A   studies.

Revised   These **results** are similar to **those of** previous studies.
Example 3-31B

Incomplete comparisons may confuse readers because their intended meaning is unclear. Complete comparisons must include both, the item being compared and the item it is being compared with.

👎	**Example 3-32**	RNA isolation is <u>more difficult</u>.

👍	**Revised Example 3-32**	RNA isolation is **more difficult than DNA isolation**.

The overuse of "*compared to*" is equally confusing for readers. Use "than" not "compared to" for comparative terms such as "smaller," "higher," or "more."

👎	**Example 3-33**	We found <u>more</u> fertilized eggs in buffer A <u>compared to</u> buffer B.

👍	**Revised Example 3-33**	We found **more** fertilized eggs in buffer A **than in** buffer B.

WRITING PRINCIPLE 22:

Avoid errors in spelling, grammar, and punctuation.

Common errors in writing include (1) spelling, (2) punctuation, (3) words that are omitted, (4) the subject and verb do not make sense together, (5) the subject and verb do not agree, and (6) unclear modifiers. When one of these errors appears, the reader is slowed down and may even need to reread the sentence to figure out the intended meaning. These errors can be avoided by carefully proofreading and double-checking the manuscript. Note that the spell check function in your word processing program has only limited applicability because many words such as *to, two, too* will not be caught when they are misspelled or chosen incorrectly.

Many scientific authors are confused about such words as "data," "spectra," and "media." "Data," "spectra," and "media" are the plural forms of "datum," "spectrum," and "medium" and thus should be treated as a plural noun. (Note that some dictionaries do accept use of both singular verbs or plural verbs with these words.)

👎	**Example 3-34**	Our <u>data suggests</u> that Klein's hypothesis is correct.

👍	**Revised Example 3-34**	Our **data suggest** that Klein's hypothesis is correct.

Another common mistake is dangling or misplaced modifiers.

Example 3-35 | <u>Having tested positive for HIV, we</u> disqualified the patients for participation in the study.

This modifier is unclear because it modifies "we". It sounds as if "we" tested positive for HIV.

Revised Example 3-35A | **Having tested positive for HIV, the patients** were disqualified for participation in the study.

Revised Example 3-35B | **Patients that tested positive for HIV** were disqualified for participation in the study.

SUMMARY OF WRITING PRINCIPLES, PART I—STYLE

WRITING PRINCIPLES—STYLE

1. Write with the reader in mind.
2. Use precise words.
3. Use simple words.
4. Omit unnecessary words and phrases.
5. Avoid too many abbreviations.
6. Use correct terminology and nomenclature.
7. Establish importance.
8. Place old, familiar, and short information at the beginning of a sentence.
9. Place new, complex, or long information at the end of a sentence.
10. Get to the subject of the main sentence quickly.
11. Place the verb immediately after the subject and the object right after the verb.
12. Use the first person.
13. Use the active voice.
14. Use past tense for observations, completed actions, and specific conclusions.
15. Use the present tense for generalizations and statements of general validity
16. Write short sentences. Aim for one main idea in a sentence.
17. Use active verbs.
18. Avoid noun clusters.
19. Use clear pronouns.
20. Use correct parallel form.
21. Avoid faulty comparisons.
22. Avoid errors in spelling, punctuation, and grammar.

PROBLEMS

Problem 3-1 Precise Words

Find the nonspecific terms in the following sentences. Replace the nonspecific choices with more precise terms or phrases. Note that it is not necessary to change the sentence structure, just replace the individual words. Guess or invent something if you have to.

1. Plants were kept in the cold overnight.
2. Some of the discovered exoplanets have an orbital period of less than 5 days.
3. The afterglow of the glowing blast wave was markedly brighter than we expected.

Problem 3-2 Simple Words

Improve the word choice in the following examples by replacing the underlined terms or phrases with simpler word choices. Do not change the sentence structure, just change the words.

1. These data <u>substantiate</u> our hypothesis.
2. We <u>utilized</u> UV light to induce the plants for mutations.
3. The differences in our results compared to those of Reuter et al. (1995) <u>can be accounted for by the fact that</u> different conditions were used.
4. <u>For the purpose of</u> examining cell migration, we dissected mouse brains.
5. <u>An example of this is the fact that</u> branching ratios differ substantially.

Problem 3-3 Redundancies

Improve the word choice of the underlined words in the following examples by removing any redundancies and unnecessary words and phrases. Do not change the sentence structure.

1. The doubling rate appeared to be <u>quite short</u>.
2. <u>It is known that</u> homologous recombination is the preferred mechanism of DNA repair in yeast.
3. Often, jewel weed <u>can be found to grow in close proximity to</u> poison ivy. (Two corrections needed.)
4. Upon heat activation, filament size increased, and the number of buds decreased. Both <u>the increase in filament length and the decrease in the number of buds</u> were only seen for cytokinin mutants.

Problem 3-4 Word Placement and Flow

Rewrite the following paragraph. Place words such that the reader can easily follow the logic flow of the message.

Mangrove plantations attenuate tsunami-induced waves and protect shore-lines against damage. Human activities on the shorelines most damaged by the great 2004 tsunami had reduced the area of mangroves by 26%. Communities can be buffered from future tsunami events by conserving or replanting coastal mangroves and greenbelts. The conservation of dune ecosystems or green belts of other tree species could fulfill the same buffer as mangroves elsewhere.

Problem 3-5 Word Placement and Flow

Write a paragraph using the list of facts provided. To create good flow, place words carefully at the beginning and end positions of sentences.

- fleas transmit plague *bacillus* to humans
- *bacilli* migrate from bite site to lymph nodes
- name "bubonic plague" arises because buboes = enlarged nodes

Problem 3-6 Word Placement and Flow

1. **Construct a paragraph about thermophiles using the list of facts provided. Pay attention to good flow of the message by considering word placement.**
2. **What does the reader expect to see next after having read the last sentence of your paragraph?**

Thermophiles
- microorganisms
- temperature range for growth between 45°C and 70°C
- found in hot sulfur springs
- cannot grow at body temperature
- not involved in infectious diseases of humans
- mechanism to resist elevated temperature unclear

Problem 3-7 Active Verbs

Put the action in the verbs in the following sentences.

1. An increase in transplant rejection occurred.
2. Two measurements of amyloid plaques were obtained for each brain.
3. Our results showed protection of the dogs by the vaccine.
4. To determine whether cell migration is occurring, we dissected mouse brains.
5. Buffalo, elephant, and black rhino abundance all show a rapid de-cline after 1977.

Problem 3-8 Pronouns

In the following sentences, the pronoun (underlined) could refer to more than one noun. Revise these sentences to make the meaning clear either by

restating the noun, by changing the sentence structure, or by repeating one or more words from the previous sentence.

1. A few microorganisms such as *Mycobacterium tuberculosis* are resistant to phagocytic digestion. <u>This</u> is one reason why tuberculosis is difficult to cure.
2. Results indicate that binding decreased about 4-fold when the temperature increased from 4°C to 17°C. (Fig. 1). <u>This</u> suggests that the binding among the particles may be governed by interactions such as hydrogen bonds or van der Waals forces that weaken when temperature rises.
3. The breakthrough was achieved because new methods and concepts could be applied, which had been developed in nonlinear dynamics to describe the spontaneous formation of order in various disciplines. <u>This</u> in turn was only possible because the underlying laws are universal at a certain abstract level.

Problem 3-9 Parallelism and Comparisons

Correct the faulty parallelism or comparisons in the following sentences.

1. The pathogenesis observed in other cells, such as circulating monocytes, may differ from endothelial cells.
2. Diabetes can be affected both by exercise and diet.
3. We observed a peak for mutant A that was higher than the other mutants.
4. Compared to the other mammals, the male dolphin was larger.
5. 'A' stars are less numerous compared to their solar-type equivalents.

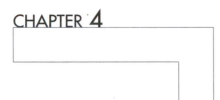

CHAPTER 4

Fundamentals of Scientific Writing, Part II—Composition

In scientific writing it is not only important to use good style but even more so to order your sentences and to structure your paragraphs and sections logically. This chapter adds to the basic scientific writing principles discussed in Chapter 3 by addressing the larger structures (paragraphs) of documents.

4.1   PARAGRAPH STRUCTURE

If paragraphs are not clearly constructed, a paper that has perfect word choice, word location, and sentence structure can be difficult to understand. The following can be considered a paragraph in which the author has not paid much attention to the needs of the reader:

Example 4-1

(1) Volcanic ash adsorption poses a great environmental hazard. **(2)** The deposition of this ash and the subsequent draining of its volatiles is a rapid route by which elements and ions are delivered to the ground (3–5). **(3)** Due to similar magma types, there appears to be some similarity in the compositions of leachates derived from volcanoes in the same regions. **(4)** The greatest hazard to the environment is posed by magmas with relatively high halogen content, and many hazardous leachate fluoride concentrations are found in volcanoes with high F/SO_4^{2-} ratios. **(5)** Finer particle sizes, <2 mm across, appear to experience enhanced adsorption, with the implication that leachate hazards may be high even where ashfall is limited (7, 8). **(6)** Enhanced growth of sulfuric acid droplets in high humidity conditions can increase gas accumulation, which increases the probability of contact with ash particles (12). **(7)** The measuring and reporting of leachate results should be standardized.

If you catch yourself reading this paragraph more than once, you are not alone. You may think that you have not paid enough attention and start reading it over. Some readers may even consider themselves not smart enough to understand the topic. The individual sentences are intelligently composed and free of grammatical errors. The sentences are also not overly long or complex. The vocabulary is professional but not beyond the scope of the educated general reader. Nonetheless, most of you arrive at the end of the paragraph without fully understanding what the author is saying. The problem lies not with you but with the author because the author has not composed the paragraph with the reader in mind. The paragraph has not been structured nor organized. Sentences seem to be put together at random and not in any logical order. Important information, such as transitions and logical connections, seem to have been left out.

Let us look at the revised version of Example 6-1:

Revised Example 4-1A

(1) Volcanic <u>ash adsorption</u> poses a great <u>environmental hazard</u>. **(2)** The <u>deposition of this ash</u> and the subsequent leaching of its volatiles is a rapid route by which elements and ions are delivered to the ground (3–5). **(3′)** <u>Adsorption</u> can be influenced by <u>magma type</u>, <u>particle size</u>, and <u>humidity conditions</u> (7). **(3)** For example, there appears to be some similarity in the compositions of <u>leachates</u> derived from volcanoes in the same regions due to similar <u>magma types</u>. **(4)** *In fact*, the greatest <u>hazard to the environment</u> is posed by <u>magmas</u> with relatively high halogen content, and many hazardous leachate fluoride concentrations are found in volcanoes with high F/SO_4^{2-} ratios. **(5)** *Aside from magma type*, <u>finer particle sizes</u>, <2 mm across, appear to experience enhanced <u>adsorption</u>, with the implication that <u>leachate hazards</u> may be high even where <u>ash deposition</u> is limited (7, 8). **(6)** *Furthermore*, enhanced growth of sulphuric acid droplets in <u>high humidity conditions</u> can increase <u>adsorption</u>, which increases the probability of contact with ash particles (12). **(7)** *Ideally*, the measuring and reporting of <u>leachate</u> results should be standardized.

After looking at this revision, you may now recognize the lack of structure and organization in the original paragraph. You can see that an important missing link was sentence 3'. It ties sentences 3 through 6 together and links them to the beginning of the paragraph. Adding the transitions "for example," "in fact," "aside from," "furthermore," and "ideally" logically connects the ideas in the paragraph. All these connections were left unarticulated in the original paragraph. Replacing "ashfall" with "ash deposition" and "gas accumulation" with "adsorption" helps to keep the reader focused on the topic because terms are used more consistently throughout the paragraph.

Although the revision greatly improved the paragraph, we could revise it even more:

**Revised
Example 4-1B**

(1') Volcanic <u>ash adsorption</u> poses a great <u>environmental hazard</u> *because <u>adsorbed volatiles</u> can be rapidly <u>deposited</u> and subsequently <u>leached</u> into the ground.* **(2')** <u>Adsorption</u> *can be influenced by <u>magma type</u>, <u>particle size</u>, and <u>humidity conditions</u>* (7). **(3)** *For example*, similar <u>magma types</u> derived from volcanoes in the same regions exhibit similar compositions of <u>leachates</u>. **(4)** *In fact*, the greatest <u>hazard to the environment</u> is posed by <u>magmas</u> with a relatively high halogen content and by <u>magmas</u> with high F/SO_4^{2-} ratios in which many hazardous leachate fluoride concentrations are found. **(5)** *Aside from magma type*, <u>finer particle sizes</u>, <2 mm across, appear to experience enhanced <u>adsorption</u>, with the implication that <u>leachate hazards</u> may be high even where <u>ash deposition</u> is limited (7, 8). **(6)** *Furthermore*, <u>high humidity</u> results in enhanced growth of sulfuric acid droplets, which increases <u>adsorption</u> by increasing the probability of contact with ash particles (12). **(7')** *Unfortunately, the use of different <u>leachate</u> analysis techniques currently prevents useful comparison between data.* **(8)** *Ideally*, the measuring and reporting of <u>leachate</u> results should be standardized.

The flow of the paragraph has been further improved in Revised Example 4-1B. Sentence 1 and 2 have been combined into sentence 1′, making their relationship clearer through the use of "because." Another link, sentence 7′, has been placed before the last sentence to more logically connect it to the rest of the paragraph.

To construct a paragraph clearly, each paragraph needs to be written such that it tells a story. Readers should be able to follow the story of each paragraph regardless of whether they understand the science. A well-constructed paragraph must not only be organized, it must also be coherent. In addition, important ideas should be emphasized, and subtopics should be signaled.

4.2   PARAGRAPH ORGANIZATION

WRITING PRINCIPLE 23:

Organize your paragraphs.

Sentences within a paragraph need to be logically organized and positioned. Each paragraph has two important power positions: the first sentence and the last sentence. Accordingly, important information should be placed in these positions rather than in the remaining sentences.

WRITING PRINCIPLE 24:

Use a topic sentence to provide an overview of the paragraph.

A well-written paragraph generally gives an overview first and then goes into detail. The topic sentence states the central topic or message of the paragraph and guides the reader into the paragraph. The end or stress position of the topic sentence highlights the topics that the author wants readers to follow in the rest of the paragraph. A topic sentence may also contain a transition from the previous paragraph or section. The details found in the remaining sentences are organized logically and consistently to explain the message provided by the topic sentence.

Example 4-2

Volatile organic compounds (VOCs) are emitted from a variety of **manmade** and **natural sources**. **Manmade sources** include motor vehicles, chemical plants, refineries, factories, consumer and commercial products, and other industrial sources. **Natural sources** responsible for biogenic VOC emissions include oak, citrus, eucalyptus, pine, spruce, maple, hickory, fir, and cottonwood. The overall relative contributions of **manmade** versus **natural sources** of VOCs have not been clearly established, but the relative contributions of these source groups vary depending on geography.

The topic of the above paragraph is "volatile organic compounds (VOCs)." The pattern of organization—that is, the order of the remaining sentences—is not random, but follows the order the items are listed: *manmade* first, then *natural*.

WRITING PRINCIPLE 25:

Use consistent order.

To keep paragraphs organized authors also need to pay attention to keeping a consistent order of topics. If you list items in a topic sentence and then describe them in the remaining sentences at the paragraph level, keep the same order.

Example 4-3

In response to a foreign macromolecule, five different immunoglobulins can be synthesized: **IgG, IgM, IgA, IgE,** or **IgD. IgG** is the main immunoglobulin in serum. **IgM** is the first class to appear following exposure to an antigen. **IgA** is the major class in external secretions, such as saliva, tears, and mucus. Thus, **IgA** serves as a first line of defense against bacterial and viral antigens. **IgA** is transported across epithelial cells from the blood side to the extracellular side by a specific receptor. **IgE** protects against parasites. The role of **IgD** is not known.

WRITING PRINCIPLE 26:

Use consistent point of view.

Be consistent in your point of view and person. Switching from one style to another within a paragraph or document disorients the reader. The point of view is consistent when the same term, or the same category term, is the *subject* of successive sentences on the same topic.

Example 4-4

This study suggests that <u>patients</u> with a prolonged febrile illness should always be a consideration for tuberculosis. Tuberculosis not only presents as fever, but <u>you</u> may also have lymphadenopathy and arthritis.

In sentences 2 and 3 the point of view is not consistent: For the second subject, the person is switched to *you*. Switches like this are very disruptive to the paragraph and disorienting for the reader.

**Revised
Example 4-4**

This study suggests that <u>patients</u> with a prolonged febrile illness should always be a consideration for tuberculosis. **Patients** with tuberculosis not only present with fever, but may also have lymphadenopathy and arthritis.

WRITING PRINCIPLE 27:

Make your sentences cohesive.

Within a paragraph, the sentences need to be cohesive so their logical layout fits neatly together. When authors arrange sentences to be cohesive, they consider word location. Good word location creates good "flow" of a paragraph.

Example 4-5

Important pathogens can be found in the genus *Yersinia*. *Yersinia* contains several **species. One species,** *Y. pestis*, is the cause of bubonic **plague**. The **plague** bacillus infects lymph nodes near the site of infection to produce buboes.

The reader conceives sentences as cohesive if the information provided at the beginning of a sentence can be found toward the end of the respective previous sentences (see Principles 8 and 9 in Chapter 3).

Placing information provided at the end of the sentences toward the beginning of the next sentence is not the only way to provide paragraph cohesion. Cohesion can also be created by providing a consistent point of view.

Example 4-6

Rhubarb is a frequently used Chinese herbal medicine. It is used to treat various ailments including constipation, inflammation, and cancer.[1,2] As a drug, **rhubarb** is made up of the roots and rhizomes of three members of the *Polygonaceae* family, *Rheum officinale, R. palmatum,* and *R. tanguticum.* Different **rhubarb** species show substantial differences in purgative effects and chemical compositions. However, **they** are similar in physical appearance and thus difficult to distinguish.

Here, information in the topic position of each sentence refers back to the topic position of the topic sentence.

Many paragraphs contain a mixture of word location. For some sentences information in the topic position may refer back to the end position of the previous sentence. For other sentences information may be written from the point of view of the old information, and refer back to the topic position of the topic sentence, as in the next example.

Example 4-7

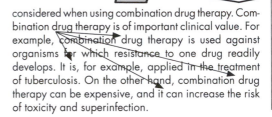

There are positive and negative factors that should be considered when using combination drug therapy. Combination drug therapy is of important clinical value. For example, combination drug therapy is used against organisms for which resistance to one drug readily develops. It is, for example, applied in the treatment of tuberculosis. On the other hand, combination drug therapy can be expensive, and it can increase the risk of toxicity and superinfection.

Constructing paragraphs in which mixed word locations occur is okay—as long as the links do not get too muddled or tangled to risk losing the reader.

WRITING PRINCIPLE 28:

Use key terms to create continuity.

Readers consider a paragraph to be coherent if they can quickly find the topic of each sentence and if they see how the topics are a related set of ideas. Thus, a coherent paragraph needs to consist of a series of sentences that lead logically from one to the next. Cohesion is achieved not only through the use of word location but also through key terms and transitions.

Key terms are words or short phrases used to identify important ideas in a sentence, a paragraph, and the paper as a whole. Usually, key terms are used to identify your main points in the topic sentence. Repeating and linking terms ensures that the topic of the work cannot be missed and that relationships between topics are clear.

To avoid confusing readers, key terms should not be changed. If you deliberately repeat key terms, your main points are emphasized and you create continuity.

Example 4-8

The dynamic binding capacity of a protein on chromatographic resins depends largely on <u>flow rate</u>. For large molecules such as proteins, the shape of the breakthrough curve may vary considerably as <u>linear velocity</u> is changed.

Especially readers unfamiliar with this particular topic may not know that "flow rate" and "linear velocity" mean the same thing. Readers may be confused when two different terms are used.

Revised Example 4-8

The dynamic binding capacity of a protein on chromatographic resins depends largely on **flow rate**. For large molecules such as proteins, the shape of the breakthrough curve may vary considerably as the **flow rate** is changed.

When you need to shift from a category term to a more specific term or the other way around, key terms should be linked so continuity is not lost and the paragraph stays coherent. To link key terms, use the category term to define the specific term.

Example 4-9

Infectious diseases related to travel may be caused by <u>gram-positive organisms</u>. **One such organism**, <u>*Staphylococcus aureus*</u>, can cause cellulitis, purulent arthritis, and suppurative lymphadenitis.

WRITING PRINCIPLE 29:

Use transitions to indicate logical relationships between sentences.

When word location and key terms alone are not sufficient to link sentences within a paragraph or section, transitions need to be used. Transitions are words, phrases, or even sentences that logically relate sentences to each other and to the overall story, such as *in addition*, *thus*, and *It is because of X*.... Transitions should be placed at the beginning of a sentence for strongest continuity and are usually set off by a comma.

Example 4-10

We determined whether the increased endotoxin susceptibility of *AUF1⁻ᐟ⁻* mice is due to deregulation of proinflammatory cytokine expression. We measured the serum TNFα level in *AUF1⁻ᐟ⁻* mice after LPS challenge.

In this example the logical relationship between sentence 1 and 2 is not clear because a transition is missing. Once the transition is added in, the relationship between the two sentences becomes obvious.

Revised Example 4-10

We determined whether the increased endotoxin susceptibility of *AUF1⁻ᐟ⁻* mice is due to deregulation of proinflammatory cytokine expression. **For this purpose**, we measured the serum TNFα level in *AUF1⁻ᐟ⁻* mice after LPS challenge.

Table 4.1 Transition Words, Phrases, and Sentences

USE	TRANSITION WORDS	TRANSITION PHRASE	TRANSITION SENTENCE
addition	again, also, further, furthermore, in addition, moreover	In addition to X, we… Besides X…	further experiments showed that…
concession	clearly, evidently, obviously, undeniably		Granted that X is…
comparison	also, in the same way, likewise, similarly, etc.	As seen in….	when A is compared with B… as reported by… when compared to…
contrast	but, however, nevertheless, nonetheless, still, yet	In contrast to A… X on the other hand… Despite X…, Unlike X..	One difference is that… Although X differed…
example	for example, specifically	To illustrate X…	An example of this is that…, that is…
explanation	here	Because of X….. In this experiment….	One reason is that…. Because X is…
purpose	for this purpose, to this end	For the purpose of…	The purpose of X was to…to determine XYZ, we…
result/	consequently, generally, hence, therefore, thus	As a result of….	Evidence for XYZ was that… Analysis of ABC showed that…
sequence/time	after, finally, first, later, last, meanwhile, next, now, second, then, while	After careful analysis of During centrifugation,…	After X was completed,… When we…
summary	in brief, in conclusion, in fact, in short, in summary	To summarize (our results),…	
Strength of transition			

4.3   CONDENSING

WRITING PRINCIPLE 30:

Make your writing concise.

A well-written paragraph should be concise. If a passage must be condensed, this often needs to be done in combination with other techniques:

1. Emphasize important information.
2. Deemphasize or omit less important information.
3. Replace or omit words and phrases.

Establishing Importance

As a first step in condensing, you need to decide what is important, what is less important, and what is unimportant information. The next steps are to emphasize the important information, deemphasize the less important information, and omit the unimportant information. Important information needs to be highlighted by placing it into a power position and/or by signaling it, such as "Most importantly, . . ." or "The key finding of this study was . . . ." Less important information can be deemphasized by subordinating it. This can be done, for example, by placing it in a subordinate clause as in the example that follows. Unimportant information adds nothing but clutter and distracts the reader. It should be omitted. Deemphasizing or omitting less important information is probably the most important technique in condensing.

Example 4-11	**Although** information about the relation between WBC count and hospital case fatality rates is limited, a link between WBC count and increased long-term mortality after acute myocardial infarction has been suggested.

Words and Phrases that Can Be Omitted

1. Omit overview words and phrases such as

described	noted
noticed	observed
reported	seen

Example 4-12	<u>Jones et al. showed that</u> intracellular calcium is released when adipocytes are stimulated with insulin. *(15 words)*

Revised Example 4-12	Intracellular calcium is released when adipocytes are stimulated with insulin (Jones et al., 1996). *(10 words)*

2. Omit phrases and sentences that tell your reader what a sentence/paragraph is about.

Example 4-13	Products were verified by gel electrophoresis. <u>The results are presented in</u> Figure 1. *(13 words)*

Revised   Products were verified by gel electrophoresis **(Fig. 1)**.
Example 4-13   *(6 words)*

3. Omit "It...that" phrases. Most of these phrases are pointless fillers and can be omitted entirely. If the idea in the phrase is essential, replace the phrase with a shorter version.

Examples of "It . . . that" phrases:

It is interesting to note that . . .	**omit**
In light of the fact that . . .	**replace (because)**
It is possible that . . .	**reword (. . . may . . .)**
It has been reported that . . .	**omit or replace (Taylor reported that . . ., or (reference))**

4. Change negative to positive expressions. This change usually results in shorter sentences. Moreover, readers prefer to read positive things, not negative ones. Above all, avoid double negatives, which can easily confuse readers.

Examples of changing from negative to positive:

negative	positive
do not overlook	note
not different	similar
not many	few
not the same	different

5. Omit excessive detail. Detail that can be inferred or is unimportant should be omitted.

Example 4-14   Using a 1-ml tip, 750 μl of the aliquot were removed into a new Eppendorf tube.

Revised   750 μl of the aliquot were removed.
Example 4-14

6. Do not overuse intensifiers or hedges.

Hedges are cautious adjectives, adverbs, or verbs such as:

usually, often, possibly, perhaps, some, most, many,
appear, could, indicate, may, seem, suggest

Intensifiers are adjectives, adverbs, or verbs that are used to strengthen nouns or verbs, such as:

always, clearly, certainly, basic, central,
crucial, show, prove

| **Example 4-15** | Figure 5 <u>clearly</u> shows that the protein was absent in the fraction. |

| **Revised Example 4-15** | Figure 5 shows that the protein was absent in the fraction. |

Within lab reports, scientific articles, and posters, intensifiers or hedges are usually avoided as data should be presented objectively. However, in proposals and job applications, intensifiers are common because these documents try to persuade rather than present.

Example 4-16	***Within a CV/resumé:***
	• Outstanding computer skills
	• Pivotal role in ABC study
	• Proven leadership ability

SUMMARY OF WRITING PRINCIPLES, PART II—COMPOSITIONS

WRITING PRINCIPLES—COMPOSITION
23. Organize your paragraphs.
24. Use a topic sentence to provide an overview of the paragraph.
25. Use consistent order.
26. Use consistent point of view.
27. Make your sentences cohesive.
28. Use key terms to create continuity.
29. Use transitions to indicate logical relationships between sentences.
30. Make your writing concise.

PROBLEMS

Problem 4-1 Paragraph Organization

The following paragraph is about the migration of salmon. Although sentences 1 and 2 describe the two possible methods salmon may employ to find their way, the paragraph does not have a topic sentence. Write a clear topic sentence for this paragraph. The topic sentence should state the message of the paragraph (the different methods salmon use to find their way during their homeward journey in the fall). In your topic sentence, make the topic the subject of the sentence. Also, make sentence 2 parallel to sentence 1.

(1) Salmon may use a magnetic or sun compass to orient themselves. (2) As described by Brown et al. (15), olfactory cues learned as smelts may also help salmon to find the river and tributary of their birth. (3) Salmon may reenter fresh water in spring, summer, or fall, but spawning occurs in the fall, and the life cycle of the salmon begins anew.

Problem 4-2 Paragraph Construction

Construct a paragraph on cancer cells using the list of facts in the order provided. In your writing, pay attention to writing a good topic sentence and to using good word location. Employ paragraph consistency, key terms, and transitions. Consider also other writing principles such as parallel form and correct pronouns and prepositions.

CANCER CELLS

- Are malignant tumor cells
- Differ from normal cells in three ways:
    - They dedifferentiate—for example: ciliated cells in the bronchi lose their cilia
    - Metastasis is possible—travel to other parts of body. New tumor growth
    - Rapid division—do not stick to each other as firmly as do normal cells

Problem 4-3 Constructing a Paragraph

Compose a short passage using the following bullet points. In this passage, create good flow by considering word location, key terms and transitions.

- The active ingredient in most chemical-based mosquito repellents is DEET (*N, N*-diethyl-meta-toluamide).
- Recent research suggests that DEET products, used sparingly for brief periods, are relatively safe (2).
- DEET is absorbed readily into the skin and should be used with caution.
- Common side effects to DEET-based products: rash, swelling, itching, and eye irritation, often due to overapplication (4).
- DEET alternatives: eucalyptus oil containing cineol, 1:5 parts diluted garlic juice, soybean oil, neem oil, marigolds, Avon skin-so-soft bath oil mixed 1:1 with rubbing alcohol, juice, and extract of Thai lemon grass (7, 8).
- DEET was developed by the U.S. military in the 1940s.
- Some research points to toxic encephalopathy associated with use of DEET insect repellents (5).
- True citronella has not been proven to be a very effective mosquito repellent (3).

Problem 4-4 Constructing a Paragraph

Use the following bullet points to create two meaningful paragraphs on polar ice reduction. In composing a cohesive paragraphs, create good flow by considering word location, key terms, and transitions.

- Sea level is projected to rise between 13 and 94 cm over the next 100 yr.
- There is continued climate warming.
- Controversial geologic evidence suggests that current polar ice sheets have been eliminated or greatly reduced during previous Pleistocene interglacials.
- Modern polar ice sheets have become unstable within the natural range of interglacial climates.
- Sea level may have been more than 20 m higher than today during a presumably very warm interglacial about 400 ka during marine isotope stage 11.
- Conflicting evidence for warmer conditions and higher sea level during marine isotope stage 11.
- Microfossil and isotopic data from marine sediments of the Cariaco Basin support the interpretation that global sea level was 10 to 20 m higher than today during marine isotope stage 11.
- The West Antarctic ice sheet and the Greenland ice sheet were absent or greatly reduced during marine isotope stage 11.
- Warm marine isotope stage 11 interglacial climate with sea level as high as or above modern sea level lasted for 25 to 30 ky.
- Variations in Earth's orbit around the sun are considered to be a primary external force driving glacial–interglacial cycles.
- Anthropogenic climate warming will accelerate the natural process toward reduction in polar ice sheets.
- Increased rates of sea level rise related to polar ice sheet decay are a potential natural hazard.

Problem 4-5 Condensing

Condense the following paragraph. Try to make your revised paragraph less than 35 words.

Our results indicate that between 5°C and 25°C, undoped, high-quality diamond as well as diamond covered with chemically bound hydrogen show no conductivity. Undoped, high-quality diamond also shows no conductivity at higher temperatures. However, diamond covered with chemically bound hydrogen clearly shows a pronounced conductivity at temperatures between 5°C and 25°C.

(51 words)

PART TWO

Working with Data

CHAPTER 5

Data, Figures, and Tables

5.1 GENERAL GUIDELINES

ILLUSTRATION GUIDELINE 1:

Decide whether to present data in an illustration or in the text.

Illustrations, such as figures, tables, photographs, and diagrams, are important components of most scientific documents and presentations. They are meant to demonstrate evidence visually. However, designing them is usually more time-consuming than describing results in the text. Therefore, decide first on the best way to present your data— that is, in an illustration or in the text.

For example, if your proposed table has only one or two rows of data, consider presenting your findings in one or two sentences in the text instead of constructing a table. Similarly, if your table lists descriptions in words rather than numbers, consider whether you really need a table—a few sentences in the text may be better. The table in the next example could easily be converted to text:

Example 5-1

Table 5.1 Antibiotic Targeting of Various Organisms

ANTIBIOTICS	ORGANISMS	CELLULAR TARGET
I	*Bacillus*	ribosomes
I	*Saccharomyces*	mitochondria
II	*Bacillus*	ribosomes
III	*Streptococcus*	cell walls
III	*Saccharomyces*	mitochondria

Revised
Example 5-1

Bacillus ribosomes were targeted by antibiotics I and II, *Streptococcus* cell walls were targeted by antibiotic III, and *Saccharomyces* mitochondria were targeted by antibiotics I and III.

ILLUSTRATION GUIDELINES 2 AND 3:

Present data in graphs when trends or relationships need to be revealed.

Prepare tables rather than graphs when it is important to give precise numbers.

When you have chosen to use an illustration rather than text, you may then need to decide what format to use for the presentation. The two most common formats in the sciences are figures and tables. It helps if you spread your data out on the table and arrange them in all possible combinations. Look for patterns. Go with a simple pattern if possible.

Example 5-2a

Table 5.2 Sample Data Presentation A

TIME (DAYS)	HORMONE A	HORMONE D54
0	200.5	455.8
5	187.1	356.7
10	166.5	321.9
15	201.1	400.6
20	289.8	500.7
25	204.1	489.9
30	189.9	389.4
35	288.9	513.4
40	205.1	499.3
45	182.9	298.5
50	278.8	533.2
55	223.4	498.5
60	199.6	250.6

Generally, choose figures when trends or relationships are more important than exact values or when hidden relationships or trends need to be revealed. Graphs are often needed to analyze data under particular statistical programs. Furthermore, when your graph includes SD or SE, it also allows readers to apprehend the existence and magnitude of any statistical differences at a glance.

Choose tables to report precise numerical information or to compare component groups, or when data are not enough to produce a satisfactory graph. Tables present data more precisely than a graph but usually do not clearly show trends within your data. Tables also present facts more concisely than text does while allowing a side-by-side comparison of the content.

Consider the next two illustrations. Which of the following presentations would you prefer given the same data?

Most readers would prefer Example 5-2b (Figure 5.1) because the trend and relationship of the data is more obvious, and exact numbers seem not as important.

Consider another example in Figure 5.2. Example 5-3 would best be depicted in the text because there are only two data points, and exact values seem important.

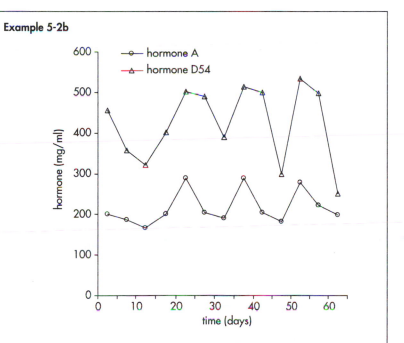

Example 5-2b

Figure 5.1 Sample data presentation B.
The same data as in Example 5-2a are presented as a line graph instead of a table.

Example 5-3

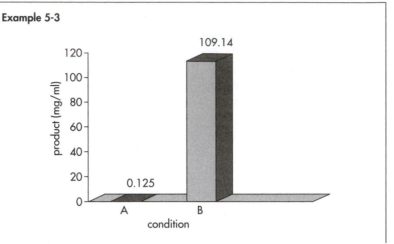

Figure 5.2 Bar graph.

5.2   FROM DATA TO RESULTS

ILLUSTRATION GUIDELINES 4 AND 5:

Prepare figures and tables with the reader in mind. Design figures and tables to have strong visual impact.

In the same way that text should be organized for the reader, figures and tables should be formatted for the reader. Figures and tables must be clear enough for the reader to get the message immediately. They must also be visually appealing while accurately showing what the text states. For example, if the text describes a newly discovered organism, the important features of this organism should be immediately visible in the figure. Similarly, if the text says that when X was done, Y increased, then in the figure, Y should look as if it increased. If the increase is not obvious, the figure is unconvincing. In this case, consider, for example, changing the scale of the axis to help illustrate the pattern, or reevaluate your interpretation of the data.

ILLUSTRATION GUIDELINE 6:

Figures and tables must be able to stand on their own.

Figures and their legends, as well as tables and their titles, must be independent of the text and of each other. Readers must be able to understand what is being portrayed without searching through the text for an explanation. Figure 5.3 and its legend, for example, stand on their own. Together they provide sufficient information for the reader to understand what is being shown. If the legend or parts of it were missing, this would not be the case.

Example 5-4

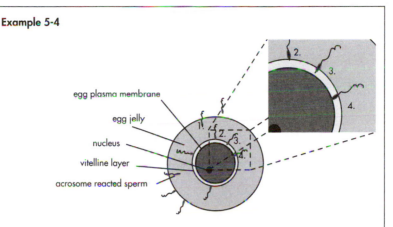

egg plasma membrane

egg jelly

nucleus

vitelline layer

acrosome reacted sperm

Figure 5.3 Diagram of *S. purpuratus* egg and sperm fertilization.
1. The egg triggers the sperm acrosome reaction. 2. The sperm
attaches the egg vitellin layer at the acrosomal process. 3. The
acrosomal process extends as it penetrates the vitellin layer. 4. The
plasma membranes of the egg and sperm fuse.

ILLUSTRATION GUIDELINE 7:

Place information where the reader expects to find it.

Most readers can understand the intended meaning of what is presented
only if the illustration has been formatted for this interpretation.

Scientists can present data in different formats. See Example 5.5,
including Table 5.3–5.5, for various possibilities for a set of data.

Example 5-5 a

0°C, 0.011% hermaphrodites	6°C, 0.011% herm.
25°C, 51/1000	18°C, 0.028%
water $T = 30°C$, 0.124%	$T = 10°C$, 2/100
35°C, 0.152%	

b

Table 5.3 Sample Data Display 1

% HERMAPHRODITES, $N = 120$	WATER TEMPERATURE (°C)
0.011	0
0.011	6
0.020	10
0.028	18
0.051	25
0.124	30
0.152	35

c

Table 5.4 Sample Data Display 2

Water temperature (°C):	0	6	10	18	25	30	35
% Hermaphrodites:	0.011	0.011	0.020	0.028	.051	.124	.152

d

Table 5.5 Sample Data Display 3

WATER TEMPERATURE (°C)	% HERMAPHRODITES
0	0.011
6	0.011
10	0.020
18	0.028
25	0.051
30	0.124
35	0.152

Although the exact same information appears in all formats, most readers prefer Example 5-5d because it is the easiest to interpret. The reason for the easier interpretation is twofold: (a) The data are written as a table, and, even more important, (b) this table is structured such that the familiar context (temperature) appears on the left, whereas the interesting results appear on the right in a less obvious pattern.

Usually, information in a table is more readily available for readers than information in the text or information presented as raw data as in Example 5-5a. However, some tables are easier to interpret than others. Generally, readers find tables much harder to follow if they are presented as shown in Example 5-5b, or if the table is horizontally arranged as in Example 5-5c. That is, we interpret information more easily if it is placed where readers expect to find it.

5.3   FIGURES, THEIR LEGENDS, AND THEIR TITLES

When you prepare a figure, consider what kind of figure you need. You can choose to present your data in a **photograph**, a **drawing**, a **diagram**, or a **graph**. Most common in the sciences are graphs, which can be presented as line graphs, bar graphs, or scatter plots.

ILLUSTRATION GUIDELINES 8–10:

Use line graphs for dynamic comparisons.

Use scatter plots to find a correlation for a collection of data.

Use bar graphs when the findings can be subdivided and compared.

Line Graphs

Line graphs, the most popular type of graph in scientific papers, use points connected by lines to show how something changes in value. Line graphs plot continuous data as points and then join them with a line. Continuous data are usually associated with some sort of physical measurement—typically, data that can fall anywhere within a range and that take on values that are fractions or decimals. In a line graph, multiple data sets can be graphed together, but a key must be used.

Do not place too much information into one figure—it will appear too packed otherwise. Do not leave much white space either—the graph will appear not well-constructed, and editors are not fond of too much open space that could be used otherwise. Three or four curves should be the maximum in a line graph, especially if the lines cross each other two or three times. When curves must cross, show which lines run where by making them of different thickness or different patterns. Draw straight lines between data points or use best-fit, smoothed curves. For an example of a line graph, see Example 5–8.

Bar Graphs

Another common graph in science is the bar graph, which uses parallel *bars* of varying lengths to display comparative values (see Example 5-6, Figure 5.4). Bar graphs tend to be more effective than line graphs for general audiences. If you use a bar graph, use a vertical rather than a horizontal bar graph because most readers are accustomed to the former. Use bar graphs in preference to line graphs for discrete data (data that can be counted in whole numbers) or when the findings can be subdivided and compared in different ways. Make the bars the same width, and make the space between bars at least one-half the bar width.

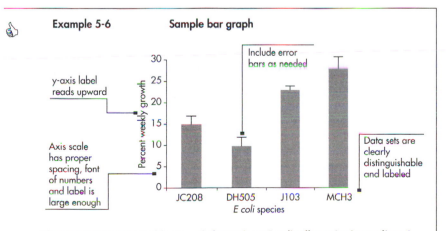

Figure 5.4 Percent weekly growth for various *E. coli* cell species in medium A.

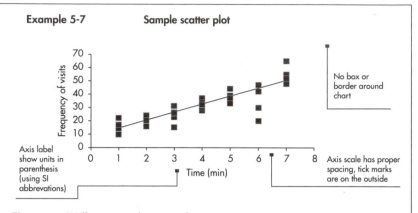

Figure 5.5 Well-presented scatter plot.
Data stand out clearly, and axes are designed and labeled well.

Scatter Plot

In a scatter plot a collection of data points is plotted using horizontal and vertical axis, similar to a line graph (see Example 5-7, Figure 5.5). Unlike in a line graph, the data points are not connected by a line. Instead, a best-fit line or trend line is added to show how the two variables (x and y) are related to each other. A corresponding equation for the correlation between the variables can be determined by established best-fit procedures. Scatter plots can be produced in two or in three dimensions.

Data points in scatter plots often overlap. That is fine. A scatter plot allows you to find a clear correlation for the points, such as a linear relationship as shown in this example.

ILLUSTRATION GUIDELINE 11:

Place the independent variable on the x axis and the dependent variable on the y axis.

Readers expect to see information in figures at certain places. For line graphs and bar graphs, readers expect to find the independent variable at the x (or horizontal) axis and the dependent variable at the y (vertical) axis. If information is placed as expected, readers can follow it easier and do not have to spend extra time trying to understand illustrations.

The important information (the data) should be immediately recognizable. Information will be much easier to understand if you emphasize it by using different line weights. For example, in line graphs, curves should be the darkest lines; letters in axis labels should be less dark; and axes, tick marks, error bars, keys, and curve labels should be least dark. Plotted points should be clearly discernable.

Example 5-8   **Sample line graph**

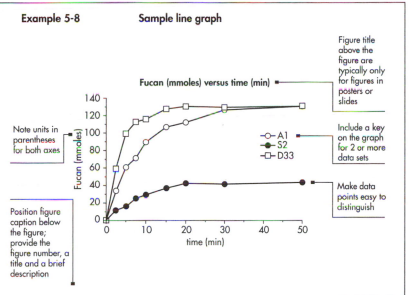

Note units in parentheses for both axes

Figure title above the figure are typically only for figures in posters or slides

Include a key on the graph for 2 or more data sets

Make data points easy to distinguish

Position figure caption below the figure; provide the figure number, a title and a brief description

Figure 5.6 Well-organized line graph.
Curves and data points are easily distinguishable. Data stand out well, and axis are designed and labeled clearly.

Table 5.6 Differences Between Figures and Tables in Text, Slides, and Posters

	TITLE	LEGEND
Figure in text	—	✓
Figure on slide	✓	—
Figure on poster	✓	✓
Table in text, on slide or poster	✓	—

Draft short but informative descriptions for the axes. Use the same symbols when the same entities occur in several figures, and use the same coordinates for different figures if values in them are to be compared.

An example of a well-constructed line graph is shown in Example 5-8 (Figure 5.6).

Although some lab reports may require you to add a title to your figure, professional scientific writing distinguishes more clearly between the requirements for figures that will appear in print and those for oral presentations or posters (see Table 5.6 and also Chapters 12 and 13).

If you are required to add a title to your figure, the easiest way to do so is by describing what is graphed in terms of y axis versus x axis as in Example 5.6; for more advanced writing, a title for the figure is usually moved into the figure legend. Often this title is somewhat more descriptive (for example: "Aggregation of fucan for different species of sea urchins" instead of "Graph of fucan aggregation versus time"). Because graphs are usually the result of an experimental question, a title for a graph could also be the answer to the experimental question (for example: "Fucan aggregation is regulated by A1 and D3").

ILLUSTRATION GUIDELINE 12:

The figure legend should be a description of the figure content.

A figure legend is a descriptive statement that is printed below or next to a figure. A legend is needed so that the figure can stand on its own without reference to the text. The legend should be an objective description of the contents of the figure.

A good figure legend typically contains

- A brief title
- Description of figure contents
- Definitions of symbols, line, or bar patterns (written in telegraph style)
- Abbreviations not defined earlier
- Possibly statistical information
- Possibly experimental details

Make sure you follow instructions given to you for your figures and their legends. Some instructors request only a title, whereas others want complete experimental details in the legend and not in the Methods section.

The figure title is the first item in the figure legend. The title should be brief and should identify the specific topic or the point of the figure (see Example 5-6 or 5-7). It should be written as an incomplete sentence and use the same key terms as are used on the graph and in the text of the paper. It should not contain abbreviations.

The descriptive material in a legend may include letters or symbols to explain special abbreviations and symbols shown in the figure. Abbreviations and symbols should be consistent between the figure and its legend, consistent between legends, and consistent with the text as well. Do not refer to the text for explanation. Note the concise listing of lanes—in what I call telegraph style—in the following example.

Example 5-9

Figure 9 Western blot of protein Z deletion analogs. Lane 1 and 10, molecular weight standards. Lane 2, protein Z. Lanes 3–6, ammonium sulfate pellets of the processed deletion analogs that exhibited agglutination. Lane 3, N18; lane 4, N74; lane 5, C166; lane 5, C121. Arrows indicate position of protein Z and its deletion analogs.

Example 5-10

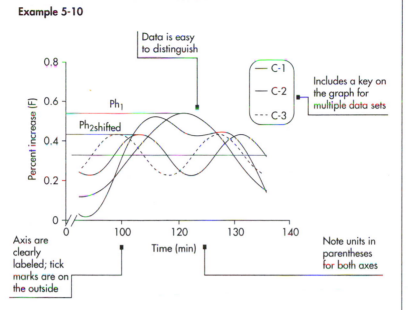

Figure 5.7 Clearly designed graph.
Curves are easy to distinguish, the graph is uncluttered, and axes are
easy to read.

ILLUSTRATION GUIDELINE 13:

Make each figure easy to read.

Differentiate points, lines, and curves well.

Label axes and scales well.

For good first impressions, make each figure easy to read (see Example
5-10, Figure 5.7). The lettering should be large enough to be legible *after*
the graph is reduced. Use the font type Arial or Arial narrow no smaller
than size 8 to allow readability.

The easiest data point symbols to distinguish are filled and open circles.
If you need three or four symbols, use filled and open triangles, squares,
and so forth in addition to the circles. Do not use X, +, 0, or *—these symbols are not distinctive enough.

Do not clutter your graphs. Avoid the use of grid lines within the
graphs. Grid lines are useful when plotting points but only rarely after-
ward. Do not extend axis lines beyond the last marked scale point, and do
not end them with an arrow pointing away from zero.

Where appropriate, draw a vertical line to show the standard deviation
(σ_x, see Section 5.6 below), the standard error (the standard deviation of
the sample mean $\bar{x}$; standard error = standard deviation/$\sqrt{(sample\_size)}$),
or a certain confidence interval (e.g., a 95% interval; see Section 5.6
below) for each data point. These lines are usually drawn in pairs in line

graphs, one above and one below a data point. For bar graphs, it is really only necessary to draw the top line of each pair. Make the lines of the error bars thinner than other lines in the main body of the data, and do not let them overlap. In the legend, tell readers what the vertical line represents, and state how many observations each mean is based on.

Log Graphs

When your data cover a large range over many powers of 10, it makes sense to graph it using logarithmic scale. This approach creates a more uniform distribution of data points. In log base 10 each successive whole number is 10 times larger than the preceding whole number.

Semilog graphs allow you to graph exponential data without having to translate your data into logarithms. As shown in Example 5-11, you could label the line across the bottom as equal to one, and the next horizontal line labeled as 1 should then be ten; then, the next 1 is 100 and the top line is 1000. You could also start with 0.001, then the next 1 would be 0.01, then 0.1 and 1.0; or you could use 10^3, 10^4, 10^5, and 10^6. Important is that each cycle division differs by a factor of 10. For any values that increase exponentially with time, one can use these graphs to determine easily doubling times. Note that the horizontal distances remain the same for each of the logarithmic sections.

ILLUSTRATION GUIDELINE 14:

Do not mislead readers.

Constructing a graph is more than just plotting points. It requires that you understand the rationale behind the quantitative methods. It also requires that you are as objective and honest as possible. Remember that the data you present may be interpreted in more than one way. You can highlight,

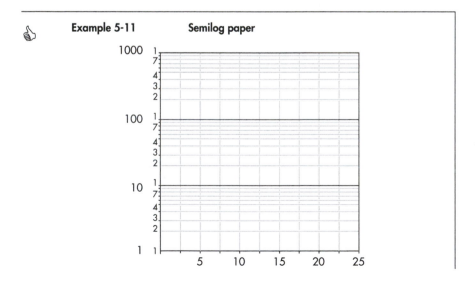

Example 5-11   Semilog paper

exaggerate, or even misrepresent a given set of data, depending on the way you choose to display it.

You should make readers aware of certain trends without misleading them. It may be very tempting to distort information and trends by, for example, deleting data points or by massaging line fits. Resist these temptations.

If a point represents the mean of a number of observations, indicate the magnitude of the variability by a vertical line centered at each point. State whether standard error or standard deviation is used, and specify the number of observations, sample size, p-values, and other statistical information. If you can, provide such statistical information in your figures and figure legends. This will lend validity to your data.

Furthermore, remember the limitations of your data. The extrapolation of a line or a curve beyond the points shown on a graph may mislead both the writer and the reader. Similarly, graphs that are to be compared but are drawn to different scales can be misleading. Draw them to the same scale and, if possible, place them side-by-side.

5.4   TABLES AND THEIR TITLES

TABLE GUIDELINES 1 AND 2:

Keep the structure of your table as simple as possible.

Place familiar context on the left and new, important information on the right.

Like figures, tables exhibit data visually, and the information presented should be independent of the text. In your tables, arrange the data to allow for easy interpretation. Because readers read from left to right, they interpret tables more easily if you place familiar information on the left and new, important information on the right. In addition, organize your tables logically to make mental comparisons easy for readers (e.g., pretreatment or control measurements should precede post-treatment or experimental measurements.) It is also easier to follow similar components when they are arranged vertically, not horizontally.

Design separate tables for separate topics and limit the number of tables. Do not use tables just to show off how much data you have collected. Also, do not repeat data in tables if you plan to put the same data in the text or in a figure, but state the most important data in the text as well.

A typical scientific table consists of a title, column headings, row (or side) headings, the body (the rows and columns containing the data), and, usually, explanatory notes. Many tables in scientific papers follow the pattern shown in Examples 5-12a and 5-12b, with row entries, column headings, and the body of the table. Within the table, make sure that your listings and numbers are aligned.

TABLE GUIDELINE 3:

Design table titles to identify the specific topic.

The titles of your tables precede the tables and identify what is being displayed. The titles have to be specific and informative without being wordy (see Examples 5-12a and 5-12b).

TABLE GUIDELINE 4:

Label dependent variables in column headings and independent variables in row headings.

Columns and rows in a table typically have a heading. Column headings should label dependent variables, and row headings should label independent variables.

Each type of information should have its own vertical column, and each column should have its own heading. Column headings consist of headings that identify the items listed in the columns below them, subheadings if necessary, and units of measurement as needed. To keep column headings brief, and thus save space in the table, use short terms or abbreviations in the column headings and subheadings, and explain the abbreviations in footnotes if necessary.

Capitalize column headings the same as the table title, but use a smaller font. Center one-line column headings, and write all headings horizontally. If headings are too long, turn them 90 degrees counterclockwise.

Row headings typically present independent variables. Keep these headings short as well, and left justify or right justify them. Capitalize the first letter of the first word, and include units where needed.

Example 5-12a   Table 5.7 General Format of a Table

Position table caption above the table and provide a title

Table 1   Initial Stocking of *R. sylvatica, A. maculatum,* and *A. opacum* in Pond Enclosures

SPECIES	POND 1 ENCLOSURE[a] (NUMBER OF INDIVIDUALS)	POND 2 ENCLOSURE (NUMBER OF INDIVIDUALS)
Rana sylvatica	100	50
Ambystoma maculatum	50	100
Ambystoma opacum	10	10

[a] Footnote a.

Headings include units in parenthesis where needed

Example 5-12b   Table 5.8 Another General Format of a Table

Table 1 Colony Diameter of *E. coli, S. aureus, and S. pneumonia* Under Different Antibiotic Media Additions

	AMPICILLIN (5 mg/l)	RIFAMPIN (10 mg/l)	STREPTOMYCIN (10 mg/l)
Diameter of *E. coli* (mm)	0.21	0.12	0.43
Diameter of *S. aureus* (mm)	0.45	0.16	0.56
Diameter of *S. pneumonia* (mm)	0.25	0.96	0

5.5   SOFTWARE RESOURCES FOR FIGURES AND TABLES

Constructing good tables and figures requires the skills of specific software programs such as Excel. Tables and figures can easily be prepared electronically. For example, you can use spreadsheets such as Excel to convert your data into a table. Alternatively, you can use Word or other word processing software to create a table directly in a word file (see also Appendix B). Note, however, that Excel is usually not sufficient to create figures and tables for professional journals and needs to be used in combination with other programs.

For the top 20 tips for using Excel, see Appendix C. Excellent tutorials for Excel are readily available online through various university sites and YouTube. Many also allow you to practice your skills at the same time. Useful web addresses for tutorials include (last accessed October 2012):

> http://office.microsoft.com/en-us/training-FX101782702.aspx. Contains links to free training courses and tutorials for diverse Microsoft products, including Excel.
>
> http://www.goodwin.edu/computer_resources/pdfs/excel_2010_tutorial.pdf. A great pdf document of an Excel 2010 tutorial with many helpful visuals.
>
> http://www.lynda.com. These Excel video tutorials teach core and advanced features and tools of Excel.
>
> http://www.gcflearnfree.org/excel2010. Contains lessons, interactives, videos and Apps for Excel 2010.
>
> http://www.excel2010tutorial.com. Lets you use a free skills book and videos to learn how to use Excel.
>
> http://www.ncsu.edu/labwrite/res/gt/gt-menu.html. Walks you through graphing with Excel.

5.6   BASICS OF STATISTICAL ANALYSIS

STATISTICAL ANALYSIS GUIDELINE 1:

Use statistical analysis to determine data trends and chance occurrence.

Experimental results often require statistical analysis to report overall trends. Knowing the most important and frequently used statistical methods to analyze findings is therefore important, particularly when you are dealing with much variation in your data. Statistical analysis allows you to judge objectively on how unusual an event is. It allows you to calculate the probability that differences as great as or greater than those observed could be due to chance. If this probability is high, then chance may be the explanation, if it is low, then you may be better able to judge general trends.

STATISTICAL ANALYSIS GUIDELINE 2:

Know the most common statistical terms and calculations.

You should be familiar with the most common terms and methods and their meaning:

mean ($\overline{x}$)   the arithmetic average of a set of values—that is the sum of the values divided by the number of the values. The mean is usually reported together with the number of measurements or sample size (n) and with the *standard deviation* or *range*.

Example 5-13
  Values: 2, 4, 8, 12, 16, 18
  $n = 6$
  $mean = (2 + 4 + 8 + 12 + 16 + 18)/6 = 10$

range   the difference between the maximum and minimum value

Example 5-14
  Values: 2, 4, 8, 12, 16, 18
  range = 18 − 2 = 16

deviation   the difference between a data point value and the mean (also the difference between observed and expected); note: deviation ≠ *standard deviation* (see below for the definition of standard deviation)

Example 5-15
  Values: 2, 4, 8, 12, 16, 18
  $n = 6$
  mean = 10

Deviation for the first data point = 2 – 10 = –8

variance (σ²) and
standard deviation (σₓ)   also known as the standard error; provides information about the variability of the data; the *variance* is calculated by summing up the squared deviations of each data value, and dividing the sum by the number of values.

Example 5-16

Values: 2, 4, 8, 12, 16, 18
$n = 6$
mean = 10
variance=$((10-2)^2+(10-4)^2+(10-8)^2+$
$(10-12)^2+(10-16)^2+(10-18)^2)/6=(64+$
36 + 4 + 4 + 36 + 64)/6 = 208/6 = 34.67

The *standard deviation* is the square root of the variance.
standard deviation = $\sqrt{34.67}$ = 5.89

95% confidence interval   derived from the standard error, indicative of a 95% chance of including the true mean.

STATISTICAL ANALYSIS GUIDELINE 3:

Understand the standardized normal distribution curve and standard deviation.

If deviations within a data set have equal variance and are normally distributed, you obtain a bell-shaped curve when you plot the values according to their frequency. This normal distribution can be standardized by making the mean equal 0 and the standard deviation equal 1. Thus, the standard deviation becomes the unit of the horizontal axis.

Example 5-17

Figure 5.8 Standardized normal distribution curve

In a standardized normal distribution, 68.3% of the data fall within −1 and 1, or one standard deviation, of the horizontal axis, and 95.4% fall within −2 and 2, or two standard deviations (see Example 5-17, Figure 5.8). Thus, the standard deviation gives a good criterion for judging the extent of error contained in the data.

STATISTICAL ANALYSIS GUIDELINE 4:

Understand the null hypothesis and statistical significance.

A null hypothesis assumes that there is no difference among two or more data sets, or between results obtained and results expected. For example, in an experiment your null hypothesis might be that there is no difference between the growth rate of chives grown at 15°C and those grown at 16°C. Although you might observe a higher growth rate for 16°C than for 15°C, there is still the possibility that your results may be purely due to chance, particularly if the variability in each sample is large. To obtain a measure of likelihood of wrongly rejecting the null hypothesis, you need to conduct particular computer statistical tests. Before conducting any parametric analysis (assuming that data has come from a probability distribution), however, you need to determine which test is the most appropriate to use based on how well your data fit certain assumptions. An overview of the most common statistical analyses is given in the following list.

Student's t-test	parametric test used to analyze quantitative data for a two-sample case in order to compare means and calculate the probability (p) that the differences among them could have arisen by chance sampling variation; if $p < 0.05$, results are generally considered *statistically significant*.
Analysis of variance *(ANOVA)*	parametric used to analyze quantitative data for two or more groups in order to compare means and calculate the probability (p) that the differences among them could have arisen by chance sampling variation
Nonparametric tests	used for quantitative data with different variation; data do not have a normal probability distribution; generally less powerful than parametric tests;
Chi-square test	used to compare observed data with data we would expect to obtain according to a specific hypothesis
Regression analysis	used to study a causal relationship between two variables, such as in a dose–response relationship

5.7   USEFUL RESOURCES FOR STATISTICAL ANALYSIS

Berry D. A., *Statistics: A Bayesian Perspective*, Wadsworth, 1996.

Bulmer M. G., *Principles of Statistics*, Dover Press, 1979.

Carlberg C., *Statistical Analysis: Microsoft Excel 2010*, Que, 2011.

Crow E. L., Davis F. A, and Maxfield M. W., *Statistics Manual*, Dover Press, 2011.

William Feller, *An Introduction to Probability Theory and Its Applications*, Wiley, 1968.

Gonick L. and Smith W., *The Cartoon Guide to Statistics*, Harper Collins, 1993.

Dan Griffiths D., *Head First Statistics*, O'Reilly Media, 2008.

Huff D. and Geis I., *How to Lie with Statistics*, W. W. Norton & Company, 1993.

Milton M., *Head First Data Analysis: A Learner's Guide to Big Numbers, Statistics, and Good Decisions*, O'Reilly Media, 2009.

Vickers A. J., *What is a p-value anyway? 34 Stories to Help You Actually Understand Statistics*, Addison Wesley, 2009.

Winston W. L., *Microsoft® Office Excel® 2010: Data Analysis and Business Modeling*, Microsoft Press, 3ed. 2011.

5.8   CHECKLIST

Use the following checklist to ensure that you have addressed all important elements for a figure or a table.

Illustrations

☐   1. Did you choose the best illustration format for your data
    ☐   a. Did you choose text if data can easily be explained there?
    ☐   b. Did you choose a line graph for dynamic comparisons?
    ☐   c. Did you choose a bar graph for findings that can be subdivided and compared?

☐   2. Is your illustration and its legend able to stand on its own?

☐   3. Did you place information where the reader expects to find it (independent variable on the *x* axis, dependent variable on the *y* axis)?

☐   4. Are your figures easy to read?
    ☐   a. Are points differentiated well?
    ☐   b. Is there a limited number of curves on your graphs?
    ☐   c. Is the lettering large enough?
    ☐   d. Has the key been added to the figure?
    ☐   e. Has a size marker been added for micrographs or other pictures?
    ☐   f. Did you label axis well and include units in parentheses?
    ☐   g. Are tick marks facing outward and are there not too many?
    ☐   h. Did you use the same scales for data that is meant to be compared?

☐   5. Did you include statistical data if needed?
☐   6. Are all abbreviations explained in the legend?

Tables

☐   1. Is the structure of your table as simple as possible?
☐   2. Did you place familiar context on the left and new, important information on the right?
☐   3. Does your table title identify the specific topic?
☐   4. Did you label dependent variables in column headings and independent variables in row headings?
☐   5. Did you include footnotes where necessary?
☐   6. Are all abbreviations explained?

SUMMARY

GENERAL ILLUSTRATION GUIDELINES:

1. Decide whether to present data in an illustration or in the text.
2. Present data in graphs when trends or relationships need to be revealed.
3. Prepare tables rather than graphs when it is important to give precise numbers.
4. Prepare figures and tables with the reader in mind.
5. Design figures and tables to have strong visual impact.
6. Figures and tables should be able to stand on their own.
7. Place information where the reader expects to find it.
8. Use line graphs for dynamic comparisons.
9. Use scatter plots to find a correlation for a collection of data.
10. Use bar graphs when the findings can be subdivided and compared.
11. Place the independent variable on the x axis and the dependent variable on the y axis.
12. The figure legend should be a description of the figure content.
13. Make each figure easy to read.
    • Differentiate points, lines, and curves well.
    • Label axes and scales well.
14. Do not mislead readers.

GUIDELINES FOR TABLES

1. Keep the structure of your table as simple as possible.
2. Place familiar context on the left and new, important information on the right.
3. Design table titles to identify the specific topic.
4. Label dependent variables in column headings and independent variables in row headings.

GUIDELINES FOR STATISTICAL ANALYSIS

1. Use statistical analysis to determine data trends and chance occurrence.
2. Know the most common statistical terms and calculations.
3. Understand the standardized normal distribution curve and standard deviation.
4. Understand the null hypothesis and statistical significance.

PROBLEMS

Problem 5-1 Figure or Table?

What format (table, graph, text) would you use for the following examples? Justify your choices.

1. You have gathered a series of data concerning ice location, thickness, and consistency in Greenland. How should you present this information?
2. You have examined mortality rates for male and female dogs exposed to the West Nile virus in all the counties of Connecticut. Should you use a table, graph, or figure?
3. You have assessed the effectiveness of a new Lyme disease vaccine after inoculating white rats with various doses of the vaccine and have measured changes in their blood pressure throughout a 2-week period. You also have a really nice photograph of one of your control rats. What should you publish and in what form?
4. You have written a paper about a new species of a bacterial pathogen implicated in a case of bacterial pneumonia. You have a chest roentgenogram showing typical findings of pneumonia and an electron micrograph of newly discovered structural details of the bacterium's flagellum. Should you include either or both in your paper?
5. You want to explain a new type of apparatus, a rotating cylindrical annulus apparatus used in your experiments. Should you use a schematic or describe it in the text or both?
6. You have constructed a new drug delivery system involving the use of nanoparticles to deliver anticancer agents to specific brain tumors. Should you explain the system using a schematic or describe it in the text?

Problem 5-2 Graph or Table

Given the raw data provided (in triplicate readings), create a table or figure, whichever represents the data best. Time is in hours, units of measurement is in ml.

$$\text{Time 0, KA} = \begin{cases} 6.7 \text{ (1. reading)} \\ 8.1 \text{ (2. reading)} \\ 7.0 \text{ (3.reading)} \end{cases} \quad \text{KB} = \begin{cases} 3.9 \\ 3.9 \\ 3.8 \end{cases} \quad \text{TFR} = \begin{cases} 0 \\ 0 \\ 0.1 \end{cases}$$

$$\text{Time 2, KA} = \begin{cases} 7.2 \\ 7.4 \\ 6.9 \end{cases} \quad \text{KB} = \begin{cases} 5.0 \\ 5.1 \\ 4.7 \end{cases} \quad \text{TFR} = \begin{cases} 1.2 \\ 2.1 \\ 1.4 \end{cases}$$

$$\text{Time 3, KA} = \begin{cases} 5.8 \\ 6.7 \\ 6.1 \end{cases} \quad \text{KB} = \begin{cases} 5.4 \\ 5.5 \\ 5.3 \end{cases} \quad \text{TFR} = \begin{cases} 1.4 \\ 1.0 \\ 1.0 \end{cases}$$

$$\text{Time 4, KA} = \begin{cases} 5.8 \\ 6.0 \\ 5.7 \end{cases} \quad \text{KB} = \begin{cases} 5.9 \\ 5.9 \\ 5.8 \end{cases} \quad \text{TFR} = \begin{cases} 2.0 \\ 1.0 \\ 1.1 \end{cases}$$

Problem 5-3 Graph or Table

Given the raw data provided, create a table or figure, whichever represents the data best.

Time 0:
> Population A: 123 members
> Population B: 54 members
> Population C: 99 members

Time 1 month later:
> Population A: 133 members
> Population B: 28 members
> Population C: 98 members

Time 2 months later:
> Population A: 142 members
> Population B: 6 members
> Population C: 98 members

Problem 5-4

Evaluate the graph in figure 5.9. How could this graph be improved? Make a list.

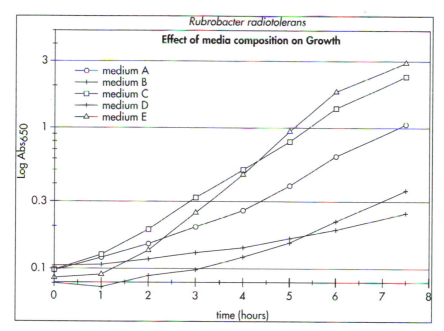

Figure 5.9

Problem 5-5

Determine the mean, variance, and standard deviation for the following data set:

Values: 12.4, 13.5, 12.7, 14.1, 12.7, 14.2, 13.8, 20.1, 15.3, 13.6, 13.9, 14.8, 12.9, 15.0, 13.8

PART THREE

Introductory Writing

CHAPTER 6

Laboratory Reports and Research Papers

6.1   OVERALL

Communicating your findings makes them available to others, and serves as an indication of your expertise and productivity. Lab reports are the most commonly encountered document in undergraduate science courses and contribute significantly to your overall grade in a lab course. However, the purpose of learning to write a good lab report is not only to let your professor evaluate you as a student. The exercise also serves as practice for writing a full scientific research article as a future professional.

It is with this foresight that this chapter describes how to write both lab reports and scientific research articles, also known as "research papers." Scientific research papers present original research findings in academic journals and should not be confused with term papers you may be asked to compose when writing a literature review in your history class. (Note that scientists also publish papers that review literature; these are known as "review articles" or "review papers" and are described in more detail in Chapter 10.) This chapter describes the basic elements of a lab report, and thus it lays the foundation for composing future scientific research papers. The chapter also discusses additional details needed for writing a full-fledged research paper. Even if you do not become a scientist, these skills are useful as you gain a better understanding of how to read and write academic publications and acquire the scientific reasoning process.

6.2   BEFORE YOU GET STARTED

Before writing your first draft, you need to understand the experiment. This includes taking good notes and being aware of the protocol to follow. It also includes writing down your observations. In addition, you have to know what format to follow to write up your report, and you may need to collect references.

GUIDELINE:
Understand the "big picture."

To write the best report, you need to understand the "big picture." To do so:

- Read your lab manual thoroughly BEFORE you start the experiment. Understand the procedure, why you are doing it, what you hope to learn, and why we benefit from this knowledge.
- Ask your instructor if you have any questions.
- Take good notes when you read your lab manual and during any lectures.
- Plan the steps of the experiment carefully. Take some time to think about the best way to organize the data before you start putting numbers down.
- Record and analyze the data carefully.
- Collaborate with your partner(s) if you are working in a group.
- When you write up your experiment in a lab report, try to target a friend not in your class. If you write with only your lab instructor in mind, you may omit material that is crucial to a complete understanding of your experiment, because you assume the instructor knows everything already.

GUIDELINE:
Obtain instructions and follow them.

Ensure that you have and understand the instructions on how to write your report. Read these instructions. Mark important details such as length of abstract, sections/headings of an article, writing style, format of references, and electronic format. Follow the instructions!

GUIDELINE:
Collect, organize, and study your references.

To write your report, you may need references. Ask your instructor for references and sources if needed. For more information on references, see also Chapter 2.

6.3   CONTENT AND ORGANIZATION

GUIDELINE:
Follow the IMRAD format.

Usually your instructor will provide you with an outline of what is expected as a lab report for the class you are taking. The overall content and organization of such a report may vary depending on your instructor and the class. Regardless of variations, however, the goal of a lab report is the same: Document your findings and communicate their significance. Ideally, your lab report should contain an Introduction, Methods, Results, and Discussion (IMRAD) section, which reflects the order of the core sections in academic journal articles/research papers:

Title page
Abstract
Introduction
Materials and Methods
Results
Discussion
Acknowledgments
References
Tables
Figures
Figure Legends

Aside from the core IMRAD sections, often in lab reports—and always in research papers—an abstract is included. Also typically included is a title page, references, figures, and tables; sometimes, the Discussion section is referred to as the "Conclusions" section, and occasionally the order of sections is changed (for example, the Methods section may appear last), depending on the discipline or audience. Overall, however, the IMRAD format represents a textual version of the scientific method: developing a hypothesis, testing it, and deciding whether your findings support the hypothesis. Reports and papers that do not follow this format, or differ significantly from it, disorient scientists. Understanding the overall format of scientific research papers and the underlying purpose of each section within them will let you adapt this overall format to the needs of a lab report for a particular course or professor.

Each section within the IMRAD format contains certain elements. To help you understand what information will go into these, Table 6.1 provides an overview of the sections. Lab instructions or journal guidelines can be used as a source of information about what to include.

Both lab reports and research papers should be centered on the overall question or hypothesis of the experiment. A good lab report or research paper also demonstrates the writer's comprehension of the concepts behind the data. Do not merely present data of expected and observed results on your report or paper. You need to show your understanding and knowledge of the principles behind the experiment and discuss how and why differences occurred. Thus, you need to organize your ideas carefully and express them coherently.

Table 6.1 Overview of Purpose and Content for Main IMRAD Sections

SECTION	MAIN PURPOSE	DETAILS
Introduction	States your overall question or hypothesis	Provides context by explaining how the hypothesis was derived and how it connects to previous research; gives the purpose of the experiment/study
Methods	Describes your experimental approach	Clarifies how you tested the hypothesis and why you performed your study in that particular way
Results	Presents the data collected	Explains how the data were evaluated; shows calculations, expresses the data in table form or as figures where needed
Discussion	Considers whether your data support the hypothesis	Interprets your results and discusses the implications of your findings; states potential limitations of your experimental design

6.4   TITLE AND TITLE PAGE

GUIDELINE:

Make the title succinct, clear, and complete.

The Title

For research papers, the title is the single most important phrase of an article because most readers will judge the paper's relevance on the title alone. You may or may not be asked to include a title for your lab report. If you need to include a title, state the main topic in 12–15 words and make the title interesting to attract readers.

A strong title should fulfill three criteria: It needs to be succinct, clear, and complete. A title is typically stated as a phrase, but it may also be a complete sentence. However, a full sentence with an active verb is usually not a good title—neither is an overly long or catchy phrase.

Do not use abbreviations in the title. The only abbreviations that are acceptable in titles are those that are better known than the words they stand for, such as DNA (deoxyribonucleic acid), and those for chemicals such as N_2O_5 (dinitrogen pentoxide). Avoid phrases such as "A Study of . . . "—these only add bulk.

Unclear titles confuse or mislead readers and can give an annoying first impression. Some titles are not clear because word choice is too general, as shown in Example 6-1.

 Example 6-1   Effect of hormones on tumor cells

This title is not clear because words in it are unspecific, such as the category terms "hormones" and "tumor cells." What hormones and which tumor cells? The specific hormones and tumor cells should be listed. Otherwise, the title is essentially meaningless. At the same time, this title is unclear because it does not state what specific effect has been observed in the study. The revised example is a much stronger title:

Revised
Example 6-1

Effect of testosterone and estradiol on the growth and morphology of rat epithelial tumor cells

Announcing the main variables of the paper is stronger than trying to fit all the variables into the title. Concentrate on the most distinctive aspect of your work. Keep in mind that in the title, readers can only absorb three or four details and that the title cannot replace the Abstract. However, be sure that your title is complete.

To check that your title is complete, compare your title with the question or hypothesis of the paper and ensure that you use the same main key terms in the title as in the question or hypothesis of the paper. Consider the following title:

Example 6-2

Isolation of *Aureobasidium pullulans*

Although this title orients the reader to the area of research, it does not give any specifics about where the organism was isolated. Adding a few more specific words completes the title and sets it apart from others in the field.

Revised
Example 6-2

Isolation of *Aureobasidium pullulans* from the vascular plant *Psilotum nudum*

The Title Page

Not all lab reports have title pages, but if your instructor wants one, a typical title page may contain the following:

- The title of the experiment
- Your name and the names of any lab partner(s)
- Your instructor's name
- The date the lab was performed or the date the report was submitted

Example 6-3   Sample Title Page

**Analysis of restriction endonuclease digests of plasmid pET3
by agarose gel electrophoresis**

Natasha Richards
Lab partner: Amy Fernández

Biology 106
Instructor: John Soders
January 25, 2011

6.5   ABSTRACT

GUIDELINE:

Abstracts of lab reports and research papers include:
Question/purpose
Experimental approach
Results
Conclusion (answer)/implication
Optional: short background and significance

The title page is usually followed by the Abstract. Knowing how to write an Abstract is one of the most important skills in science, because virtually all of a scientist's work will be judged first (and often last) based on an abstract. The ability to write a competitive abstract applies not only to lab reports but also to other critical scientific documents such as research papers, grant proposals, progress reports, project summaries, and conference submissions. Therefore, learning this critical skill cannot be underestimated.

The Abstract should fully summarize the contents of the report or paper in one paragraph. It must also be written such that it can stand on its own without the text. Include only the most important details of the paper and use as few words as possible. Abstracts for lab reports are typically 50 to 150 words long. Those of academic research papers are usually longer and range from 100 to 250 words.

Although the Abstract is a mini version of the report, it does not give equal weight to all the parts of a report. The Abstract may include a sentence or two of background information. It needs to include the overall question or purpose of the work. It typically describes the experimental approach only generally and includes only the main results from the Results section. In the Discussion, it also contains the answer to the research question/purpose. In addition, the Abstract may end with a sentence stating an implication, a speculation, or a recommendation based on the answer. It should not end, however, with a general descriptive statement that merely hints at your results.

In your Abstract, do not include any information or conclusion not covered in the paper. Avoid abbreviations, unfamiliar terms, and citations. Do not include or refer to tables or figures, nor any references, but be sure to include all the important key terms found in the title because the Abstract and the Title have to correspond to each other. You may want to wait to write your abstract until you have completed all other parts of your report because the Abstract will contain the main elements of all the other sections.

GUIDELINE:
Signal the elements of the Abstract.

Because Abstracts are usually written as one paragraph, it helps the reader if you signal the different parts of the Abstract. Examples of signals for the abstract are shown in Table 6.2.

Example 6–4 is an example of a well written abstract that contains all the necessary elements.

Table 6.2 Signals for the Abstract

QUESTION + EXPERIMENT	RESULTS	ANSWER/ CONCLUSION	IMPLICATION
To determine whether..., we...	We found...	We conclude that...	These results suggest that...
We asked whether......	Our results show...	Thus,...	These results may play a role in...
To answer this question, we...	Here we report...	These results indicate that...	Y can be used to...
X was studied by...			

👎 **Example 6-4**   **Sample Abstract**

Hypothesis   We tested the hypothesis that an increase in temperature
                    results in faster development of tadpoles into frogs. In our
Experimental   experiment, bullfrog tadpoles (*Rana catesbeiana*) were
Approach   kept at either 15°C, 16°C, 17°C, 18°C or 20°C warm
                    water, and their rate of development was evaluated by
                    scoring the appearance and growth rate of limbs over a
Results   period of four weeks. We observed that tadpoles living
                    in warmer water developed limbs faster than those living
Conclusion   in cooler water. These results suggest that tadpoles liv-
                    ing in warmer climates or warmer ponds might turn into
                    frogs sooner due to more favorable living conditions.

6.6   INTRODUCTION

GUIDELINE:

The Introduction should follow a "funnel" structure:
Background
Unknown/Problem
Question/Hypothesis
Experimental Approach
Optional: Results/Conclusion
            Significance

The purpose of the Introduction is twofold: to interest your audience to
read the paper and to provide sufficient context or background information for readers to understand why your study was performed and what
specific research question or hypothesis you addressed.

Introductions for lab reports are usually short—some are as short
as a single paragraph—stating the question/purpose or hypothesis of the
experiment and the approach. In addition, they may contain some background information.

👍 **Example 6-5**   **Sample Introduction for a lab report**

                    In this experiment we tested the effect
Purpose   of different water sources on tobacco
                    pollen germination. *Nicotiana tabacum*
Experimental   L.var. Petit Havana (tobacco) was sus-
Approach   pended in germination medium pre-
                    pared with either tab water as delivered
                    to the laboratory, pond water collected
                    from Cedar Lake pond, or deion-
                    ized water prepared in the laboratory.
                    Germination was evaluated after 24
                    hours under an inverted microscope.

In comparison to that of a lab report, the Introduction of a typical journal article/research paper is longer—typically, one to two double-spaced pages or about 250–600 words. Longer introductions, such as those for research papers, contain additional details and background information, as well as a very specific unknown/problem of interest in the respective field to explain the overall purpose of the study (see Table 6.3). Generally, readers expect the parts of the Introduction for research papers to be arranged in a standard structure: a **"funnel,"** starting broadly with background information and then narrowing to what is the problem/unknown, the question, and the experimental approach of the paper (see also Zeiger, 2000). Often, the Introduction also gives an overview of the main results and implications of the report or paper. (Note that the Introduction may repeat some parts of the Abstract, which is okay.)

Table 6.3 Guidelines for Introductions for Lab Reports and Research Papers

	LAB REPORT	SCIENTIFIC RESEARCH PAPER
Background	Some background information	Broad and specific background information and previous research in the area
Unknown/Problem	Usually not provided	Problems of previous work and unknown factors in the area
Question/Hypothesis of Study	Indicates purpose of experiment	Addition made by your research
Experimental Approach	Approach taken to perform experiment	Approach taken toward this addition

If you are writing a lab report that is more like a full scientific research paper, you will need to provide one to two paragraphs of background information. To do so, you can use lecture notes and descriptions provided in your text or laboratory book. You may also need to do more research using the Internet and library to search recent scientific literature to find other research in this area of study. Summarize that research, stating what the general findings have been and use those findings to describe the current knowledge in the area. For longer lab reports, this background serves to demonstrate that you understand the context for the experiment or study you have completed.

The most important element in the introduction is the research question/hypothesis. It should name the variables studied as well as the main features of the study. Note that the question/purpose is usually not written in the form of a question but as an infinitive phrase or as a sentence, using a present tense verb.

	Example 6-6	**Phrasing of question/purpose**
	a	**To determine** if increasing amounts of CO_2 in the atmosphere affect maize seed germination, . . .
	b	**We examined** the effects of various low temperature durations on the growth rate of the moss *P. patens*.

Following the question in the Introduction of your research paper, briefly indicate your experimental approach—usually one sentence, at most, two or three sentences. Signal the experimental approach so readers can identify it immediately.

	Example 6-7	Phrasing of the experimental approach
	a	**We analyzed** restriction enzyme digestion **by** agarose gel electrophoresis.
	b	The number of eggs per cluster **was characterized by** light microscopy.

GUIDELINE:

Signal the elements of the Introduction.

Generally, all the parts of the Introduction should be signaled so the reader does not have to guess about the information provided. The signals vary, depending on how the known, unknown, question, and experimental approach are phrased. Numerous variations on these signals are possible. Some examples are listed in Table 6.4.

Table 6.4 Signals of the Introduction

BACK-GROUND	UNKNOWN	QUESTION, HYPOTHESIS	EXPERI-MENTAL APPROACH	RESULTS	IMPLI-CATION
X is...	...is unknown	We hypothesized that...	To test this hypothesis, we...	We found...	...consistent with
X affects...	...has not been determined	To determine...	we....	...was found	...indicating that
X is a component of Y	The question remains whether....	To examine..., To assess... To analyze... In this study we examined...	We analyzed... For this purpose, we ...	We determined ...	...make it possible to ...
X is observed when Y happens...	...is unclear	Here we describe ...	... by/using ...	Our findings were...	...may be used to...
X is considered to be...		This report describes...	For this study we...	We observed that...	...is important for...
X causes Y	...does not exist ...is not known	We examined whether X is... We assessed if... In this experiment we determined if... In our experiment we	To evaluate..., we... To answer this question, we...	Based on our observations ...	Our analysis implies/ suggests... Our findings indicate that...

A good introduction, in which the background starts very broad but then narrows down quickly to the research topic, is shown in Example 6–8:

Example 6-8   **Partial introduction showing good funneling of background**

Broad background

In mammals, the auditory hair cells of the inner ear are the sensory receptors of the auditory system. Two functionally and anatomically distinct types of mammalian auditory hair cells exist: inner and outer hair cells. Outer hair cells do not send neural signals to the brain, but they mechanically amplify low-level sound that enters the inner ear (1). The amplification is powered by an electrically driven motility of their cell bodies (2).

Specific background

Unknown/Problem

The molecular basis of this mechanism is thought to be the motor protein prestin, which is embedded in the lateral membrane of the outer hair cells. Mammalian prestin is an 80-kDa, 744-amino-acid membrane protein whose function appears to depend on chloride channel signaling (3,4). Although prestin has been researched intensively, its molecular function has not been fully established.

A complete Introduction, containing all elements, is shown in Example 6–9. Note how this Introduction funnels from broad to specific background to the unknown, the question, and the experimental approach. The latter two elements are stated in the last paragraph of the Introduction, which represents an important power position of this section.

Example 6-9

Broad to specific background

Problem

The species-specific attachment of sperm to sea urchin eggs provides a model system for the biocheminal study of intercellular adhesion. The protein bindin is located in the acrososme granule of sea urchin sperm and is known to be the sperm component of the sperm–egg bond (Vacquier, 1977). Bindin proteins isolated from two different species of sea urchins show similar yet distinct amino acid compositions and sequence (Bellet et al., 1977). Further study of *S. purpuratus* species-specific bindin and its interaction with eggs requires large amounts of the protein. As isolation from sea urchins is inefficient, a new method of production had to be found.

More background

The use of bacteria for protein expression has been a powerful tool of genetic engineering. Not only does this use of bacteria allow production of proteins in biochemically useful quantities, but also the systematic alteration of the protein structure by site-directed mutagenesis of the cDNA and subsequent study of the effect of such alterations on the function of the protein *in vitro*.

Question/ purpose and experimental approach

In this study, we expressed the sea urchin protein bindin in *E. coli* in order to isolate it in large amounts for functional studies. The expression vector developed to accomplish this could also be used for future site-directed mutagenesis experiments.

6.7   MATERIALS AND METHODS

Provide enough details and references in the Materials and Methods section to enable a trained scientist to evaluate or repeat your work.

The Materials and Methods section describes the experimental approach used to arrive at your conclusions. You can organize your Materials and Methods section by separating groups of actions into subsections, each with its own subheading. Although use of subsections is optional, it usually simplifies and clarifies the presentation for the reader. The sequence of events within these subsections is then written in chronological order or from most to least important.

Example 6-10   **Materials and Methods subsection**

Cultures. Samples taken from the tips and subcutaneous sections of aseptically removed IVs were cultured as previously described (12). To identify the sources of organisms that colonize IVs, swab cultures of surrounding skin were obtained at the time of IV insertion as well as at the time of IV removal. In addition, one or more peripheral blood samples were obtained. Isolated organisms were identified by standard microbiologic methods.

The Materials and Methods section should contain sufficient detail to evaluate or repeat your work. You should write this section with care so that your experimental approach does not appear faulty, incomplete, or unprofessional.

The Materials and Methods section should cover

- Materials (drugs, culture media, buffers, gases, or apparatus used)
- Subjects (patients, experimental materials, animals, microorganisms, plants)
- Design (includes independent and dependent variables, experimental and control groups)
- Procedures (what, how, and why you did something)

Define the materials and methods as precisely as you can. You need to provide sufficient details and exact technical specifications such as temperature, pH, total volume, time, and quantities to ensure that scientists can repeat your work. Do not forget to include your control experiments.

Following are a few examples that do not provide sufficient detail:

Example 6-11   **Providing insufficient detail**

a   To classify native species, orchids were collected.

b   To identify genes with a high probability of having differential expression in follicular carcinomas, we used statistical methods.

c   All samples were centrifuged.

Revised Example 6-11

a   To classify native species, orchids were collected **in the Everglades during the month of January.**

b   To identify genes with a high probability of having differential expression in follicular carcinomas, we used **parametric (*t* test) and nonparametric (Mann–Whitney *U* test)** methods.

c   All samples were centrifuged **at 5000 x g for 30 min at 25 °C.**

Be sure not to go overboard providing too much detail, such as tube or pipette size or color, or the manufacturer's name unless this is relevant to your experiment.

Example 6-12   **Unnecessary information in Materials and Methods**

Cells were scraped out of the wells and resuspended <u>in a 1.5 ml Eppendorf tube.</u>

Revised Example 6-12

Cells were scraped out of the wells and resuspended **in 100 µl of sterile saline.**

GUIDELINE:
Provide literature references if needed.

If your methods have not been reported previously, you must provide all of the necessary detail. If, however, methods have been described previously, such as in a lab manual, you may provide only that literature reference. If you modified a previously published method, provide the literature reference and give a detailed description of your modifications.

Example 6-13   **Referring to a previously described method**

Plasmids were isolated as described in the Bio 112 lab manual.

Example 6-14	**Referring to a previously described method with modifications**
	Plasmids were isolated as described in the Bio 112 lab manual with minor modifications. Instead of dissolving DNA pellets in sterile water, pellets were dissolved in buffer A.

It is important to ensure that the reader will understand why each procedure was performed and how each procedure is linked to the central question of the paper. Therefore, you should state the purpose or give a reason for any procedure whose function or relation to the question of the paper is not clear.

Example 6-15	**Statement of purpose**
	To purify X, the mixture was run over a Y column.

The Materials and Methods section is the one section in a research paper where often passive voice is preferred over active voice. The reason is two-fold: It lets you emphasize materials or methods as the topic of your sentences, and readers do not need to know who performed the action.

Example 6-16	**Use of voice**
	The assays **were performed** for 10 min at room temperature.

In the Materials and Methods section, the general rule about verb tense applies (see Chapter 3). When reporting completed actions, use past tense. However, use present tense for statements of general validity and for those whose information is still true or if you are referring to figures and tables.

Example 6-17	**Use of tense**
a	pEG12 **was digested** with EcoRI.
b	Criteria used in selecting subjects **are listed** in Table 2.

6.8   RESULTS

Report your main findings in the first paragraph of the Results section.

Structure each experimental description in subsequent paragraphs by stating:

Purpose or background of experiment
Experimental approach
Results
Interpretation of results (for research papers)

The Results section presents the results of your experiments and points the reader to the data shown in the figures and tables. Do not forget to include control results and, if needed, explain the purpose of an experiment shortly.

Start the Results section by presenting your main findings in the first paragraph. Your main findings are the findings used in providing the overall answer/conclusion of the paper. You may also start the first paragraph with a brief overview of your general observations and then move on to the main findings. In the latter case, do not devote more than a few sentences to any overview, and ensure that your main findings still appear in the first paragraph, because it is a power position.

Example 6-18   Results—First Paragraph

Overall background/purpose and experimental approach

Main overall results

During fertilization experiments performed to determine the fertilizability of the sea urchin species *S. purpuratus*, we discovered that as many as one third of the sea urchins examined appeared to be infertile. Furthermore, 0.5% of the urchins were hermaphrodites, that is, they contained gametes, eggs, and sperm. Substantial variations in fertilization were observed for approximately 30% of the individual urchins (Fig. 2).

In subsequent paragraphs, present your specific observations. Information in these paragraphs needs to be organized. For simple lab reports, each paragraph that describes results of a specific individual experiment should contain the following essential components:

- Purpose or background of experiment if needed
- Experimental approach
- Results

Start your segments or paragraphs by providing a topic sentence. This topic sentence usually indicates the purpose of the experiment performed. It may also provide context in form of background information. The purpose is followed by a short statement of your experimental approach (about half a sentence). The purpose may be written in the form of a transitional phrase or clause, for example. Follow the experimental approach immediately with your results for the experiment. Place important or general results first and less important details later in the paragraph.

Table 6.5 Guidelines for the Results Section of Lab Reports and Research Papers

	LAB REPORT	SCIENTIFIC RESEARCH PAPER
First paragraph	Present main results	Provide an overview of main results
Subsequent paragraphs	Elements	Elements:
	• Purpose or background of experiment if needed	• Purpose or background of experiment if needed
	• Experimental approach	• Experimental approach
	• Results	• Results
		• Interpretation of results

For longer, more in-depth lab report and for research papers, a general interpretation of the results should be added to each paragraph or portion to make the results meaningful for the reader. Note that this last sentence in a results paragraph should only state what the results mean without discussing them further—a discussion will follow in the Discussion section.

In such longer lab reports or research papers, you may even consider dividing your Results section into different subsections to make navigating through the section easier for the reader. Ensure, however, that you report only results that are pertinent to the question posed in the Introduction and to the experiments described in Materials and Methods. Exclude preliminary results and results that are not relevant. Also incorporate results whether or not they support your hypothesis, and explain any contradicting results if necessary.

Example 6-19	**Well-formed Results paragraph**
Background	**1**Considerable evidence suggests that ATP is needed in the binding of mRNA to the 40S ribosomal subunits (13). **2**To understand the interaction between ATP and mRNP particles better, **3**we incubated the mRNP particles with ^{14}C ATP at optimal concentrations for *in vitro* yeast translation. **4**Results indicate that ^{14}C ATP bound to mRNP particles, but the binding decreased about 4-fold when the temperature increased from 4 to 17°C (Fig. 1). **5**These results suggest that the binding between ATP and mRNP particles may be governed by interactions such as hydrogen bonds or van der Waals that weaken when temperature rises.
Purpose	
Experimental approach	
Result	
Interpretation of results	

Regardless of whether you are composing a Results section for a lab report or one for publication in a journal, throughout the Results section, emphasize the data and their meaning. Subordinate control results and methods.

GUIDELINE:
Distinguish between data and results.

To present your results to the reader clearly, you need to distinguish between data and results. Data are values derived from scientific experiments (concentrations, absorbance, mean, percent increase). Results interpret data:

Example 6-20   Absorbance **increased** 55% when samples were incubated at 25°C instead of at 15°C.

If data is provided without interpretation or explanation, findings are not very meaningful to readers.

Example 6-21   **Presenting data without interpretation**

Among the 785 obese mice, we found 522 males and 163 females.

To interpret the data for the reader, the author needs to first present the interpretation and then the data that supports it.

Revised   We found that **3.8 times as many** male mice (79.2%)
Example 6-21   than female mice (20.8%) were obese.

When you state your interpretations/results, ensure that you make reference to your data in figures or tables by referencing the figure or table number in parentheses where needed.

Example 6-22   The calcium concentration in our unknown sample was determined to be 15 ppm **(Fig. 3)**.

Although most data should be presented in figures and tables, your main findings should be stated in the text as well along with your interpretations of all data.

GUIDELINE:
Report results in past tense.

Results are usually reported in past tense because they are events and observations that occurred in the past.

Example 6-23   **Use of tense**
          (a)   Imidazole **inhibited** the increase in arterial pressure.
          (b)   Once nectar **was depleted** from the *D. wrightii* flower, all of the moths then **switched** to feeding from the *A. palmeri* flowers.

Table 6.6 Signals for the Results

PURPOSE/QUESTION	EXPERIMENTAL APPROACH	RESULTS	INTERPRETATION OF RESULTS
To determine...	...we did...	We found...	, indicating that...
To establish if...	X was subjected to...	We observed...	, consistent with...
Z was tested...	... by/using...	We detected...	, which indicates that...
For the purpose of XYZ...	ABC was performed...	Our results	This observation
	Experiment X showed...	indicate that...	indicates that...
		that...	A is specific for...

GUIDELINE:

Signal the elements of the Results section.

As in the Abstract and Introduction, the different elements in the Result section should be signaled so that readers cannot miss them. To emphasize different elements, use signals such as those listed in Table 6.6.

6.9   DISCUSSION

GUIDELINE:

Organize the Discussion

First paragraph:	Interpretation/Answer based on key findings
	Supporting evidence
Middle paragraphs:	Secondary results
	Limitations of your study
	Unexpected findings
	For research papers also:
	Comparisons/contrasts to previous studies
	Hypotheses or Models
Last paragraph:	Summary
	Significance/Implication

The Discussion is usually the hardest section to define and to write. Even though the data may be valid and interesting, the interpretation or presentation of it in the Discussion may obscure it. Therefore, good style and clear, logical presentation is especially important here.

The main function of the Discussion is to interpret your key findings and to draw conclusions based on these findings—in other words, answer the question(s) asked in the Introduction. The discussion should also explain how you arrived at your conclusions.

In the Discussion, you should include explanations for any results that do not support the answers. In addition, you may discuss any possible errors or limitations in your methods, give explanations of unexpected findings, and indicate what the next steps might be. Do not refer to every detail of your work again. Instead, in your Discussion, summarize and generalize.

In the first paragraph, tell your readers what your key findings were and what they mean. In subsequent paragraphs, explain how your findings fit into what is known in the field. In the last paragraph, summarize and generalize why the contribution of your study is important overall, in your field, outside your field, and/or for society.

Because the interpretation of your key findings is the most important statement in the paper, it should appear in the most prominent position: the first paragraph of the Discussion. The interpretation of your key findings should match the question/purpose for the study stated in the Introduction and answer what the introduction asked. The interpretation of your key findings should also be repeated in the other power position of this section: the last paragraph.

Do not begin the Discussion with a second introduction, a summary of the results, or secondary information. Begin by directly stating the answer based on your findings in the opening sentence of the Discussion. If you feel that this beginning is too abrupt, you can restate the purpose of the study or provide a brief context before stating the answer. Any statements placed before your answer should not exceed more than a few sentences.

Example 6-24   **First paragraph of Discussion**

Question/Purpose:

Our goal was to determine **what part of the *bindin* polypeptide is responsible for the species-specific egg agglutination activities of the protein**.

Answer/Interpretation of key findings:

Answer to question | *Our results suggest that* **the part of bindin responsible for species-specific egg agglutination lies in the region of residues 75–121**. *We showed that* **residues 18–74 and 122–236 can be deleted without loss of egg agglutination activity**. All of the biologically active bindin deletion analogs were found to be **species-specific** by their ability to **agglutinate** exclusively *S. purpuratus* eggs. Deletion analogs that had any residues of region 75–121 deleted exhibited no significant activity above
Supporting evidence | the bacterial control protein.

(With permission from Elsevier)

In Example 6–24, the interpretation of the key findings matches the purpose of the study. Note that the same key terms (*bindin, species-specific, egg agglutination*) appear in the question/purpose as well as in the interpretation or answer to the question. This answer is immediately supported by key findings of the study.

After stating and supporting your answer, mention other findings that were important. Tell your readers what you think your results mean and how strongly you believe in them. Organize these findings according to the science or from most to least important. To ensure that your Discussion is organized rather than rambling, focus the story on the question/purpose of the paper that was stated in the Introduction. Treat your secondary results as you did your main findings: Summarize and generalize them rather than simply repeating what you found.

In these subsequent paragraphs, you may also mention any limitations of your study or unexpected findings and may present any new hypothesis or model based on your findings. If useful, include figures to illustrate complex models in the discussion.

Example 6-25	**Explaining limitations in the Discussion**
	Our data show that Aβ assemblies did not colocalize in drusen. It is important to note, however, that the epitope for Aβ may have been masked within the oligomeric structure, as is the case when Aβ monomers are transformed
Limitation	into amyloid fibrils (40). Therefore, we cannot preclude the possibility that the oligomeric cores in drusen are made up of Aβ.

For in-depth lab reports or research papers, compare and contrast your findings with those of previously published papers, but avoid the temptation to discuss every previous study in your subject area (see Table 6.7 for the difference in the discussion section of lab reports and research papers). Stick to the most relevant and most important studies. Explain any disagreements objectively, and credit and confirm the work of others by referencing it. Give pro and contra arguments for your conclusion. Only if you mention both impartially will you sound convincing to the reader. You may also state

Table 6.7 Guidelines for the Discussion Section of Lab Reports and Research Papers

	LAB REPORT	SCIENTIFIC RESEARCH PAPER
First paragraph	Present main results and their interpretation	Present main results and their interpretation
Subsequent paragraphs	Provide: • Secondary results and their interpretation • Limitations of your study • Unexpected findings	Provide: • Secondary results and their interpretation • Comparisons/contrasts to previous studies May also provide: • Limitations of your study • Unexpected findings • Hypotheses or models
Conclusion	Summarize main findings and indicate their importance	Summarize main findings and indicate their importance

theoretical implications or practical applications. Know that most of the time it is wise to present your opinion carefully rather than too strongly.

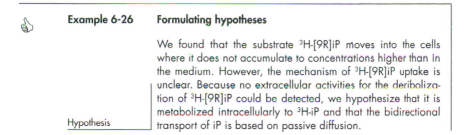

	Example 6-26	**Formulating hypotheses**
		We found that the substrate ^{3}H-[9R]iP moves into the cells where it does not accumulate to concentrations higher than in the medium. However, the mechanism of ^{3}H-[9R]iP uptake is unclear. Because no extracellular activities for the deribolization of ^{3}H-[9R]iP could be detected, we hypothesize that it is metabolized intracellularly to ^{3}H-iP and that the bidirectional
	Hypothesis	transport of iP is based on passive diffusion.

GUIDELINE:
Provide a conclusion.

At the end of the discussion, you should provide some closure by writing a one-paragraph concluding summary. Readers typically expect to see two things in the summary of a scientific paper: an analysis of the most important results and the significance of the work. The analysis of the most important results is typically provided by the interpretation of your key findings—that is, the answer. Here, too, the answer should match the question/purpose you posed in the Introduction and the answer presented in the first paragraph of the Discussion. Do not bring in new evidence for the summary. Rather, complete the "big picture" by restating your answer—that is, the interpretation of the key findings.

Table 6.8 Signals for the Discussion

ANSWER	KEY FINDINGS	SUMMARY	SIGNIFICANCE
In this study, we have shown that…	In our experiments… …can be attributed to…	In summary,… In conclusion,… Finally,…	Our findings can/will serve to… …can be used…
In this study, we found that…	We determined X by…	Taken together,…	We recommend that X is… Y should be used for… …is probably…
Our study shows that…	We found that…	To summarize our results,…	Y indicates that X might…
Our findings demonstrate that…	Our data shows that…	We conclude that… [overall question]	These findings imply that X may…
This paper describes…	…has been demonstrated by…	Overall,…	Here we propose that We hypothesize that

GUIDELINE:
Signal the different elements in the Discussion.

Signal the different elements of the discussion so that readers recognize immediately what they are reading about. Possible signals for various elements are listed in Table 6.8.

6.10   REFERENCES

In all research papers and lab reports, your interpretation of the available sources must be backed up with evidence. Therefore, cite primary and secondary sources where needed and include a reference list. The type of information you choose should relate directly to the article's focus. See Chapter 2 for more information on reference selection and citation.

6.11   ACKNOWLEDGMENTS

Lab reports usually do not contain an Acknowledgments section, but research papers do. In this section, acknowledge any organizations or individuals who provided grants, materials, and financial or technical assistance as well as those who contributed ideas, information, and advice to your work. You do not need to acknowledge those who did no more than their routine work. People named in the Acknowledgments should give their permission to be named and should approve of the wording of your acknowledgment.

For the order of acknowledgments, start by listing intellectual contributions, then move on to technical support, provision of materials, helpful discussions, and revisions and preparations of the manuscript. Last, list any funds, grants, fellowships, or financial contributions.

Example 6-27   We thank Dr. J. Holzheimer for his technical advice on crystallization assays. We are also grateful to Drs. Thomas Hugh and Fred Grant for their critical review of the manuscript. This study was supported by grant # XXX from the National Institutes of Health.

6.12   CHECKLIST

Use the following checklist to ensure that you have addressed all important elements for a lab report or research paper. In your revisions and in editing papers, work your way backward from paragraph location and structure to word choice and spelling.

Abstract
☐   1. Is the question/purpose stated?
☐   2. Is the experimental approach stated?
☐   3. Are the results indicated?
☐   4. Is the answer/conclusion provided?

☐ 5. Are all elements signaled?

☐ 6. Is the length within the required limits?

☐ 7. Is the significance of the work apparent?

Introduction

☐ 1. Are all the components there?

    ☐ a. Background

    ☐ b. Unknown

    ☐ c. Question/purpose

    ☐ d. Experimental approach

    ☐ e. Results/conclusion

    ☐ f. Significance

☐ 2. Is the research question stated precisely? (Is it in present tense?)

☐ 3. Do all the components logically follow each other? (Is the unknown what one would expect to hear after reading about what is known? Is the research question really the question one would expect to read after reading the unknown? Does the answer really answer the research question?)

☐ 4. Is all background information directly relevant to your research question? Did you list only the most pertinent literature and not review the topic?

☐ 5. Have all elements been signaled clearly?

☐ 6. Are references placed correctly and where needed? (Chapter 2)

☐ 7. Is the introduction cohesive and coherent? (see Chapter 4)

Materials and Methods

☐ 1. Do the listed materials and methods describe all procedures done to obtain the results presented?

☐ 2. Are sufficient details and/or references provided?

☐ 3. Are protocols logically grouped and organized?

☐ 4. Are topics signaled and linked?

☐ 5. Did you pay attention to voice (mainly passive)?

☐ 6. Did you ensure that major results are not stated in the Materials and Methods section?

☐ 7. Is the purpose stated for any procedure whose function is not clear?

Results

☐ 1. Did you report all main findings as well as other important findings?

☐ 2. Are your most important results and their interpretation provided in the beginning of the Results section?

☐ 3. Does each Results segment or paragraph contain all components (purpose of experiment, experimental approach, results, and their interpretation)?

    ☐ a. Is the purpose of each experiment apparent?

    ☐ b. Is the experimental approach provided?

    ☐ c. Are results interpreted?

☐   4. Are all components (purpose of experiment, experimental approach, results, and their interpretation) signaled?

☐   5. Is the reader pointed to figures and tables?

☐   6. Are control results included?

Discussion

☐   1. Did you interpret the key findings?

☐   2. Is the interpretation/answer to the research question in the first paragraph?

☐   3. Is the answer followed by supporting evidence?

☐   4. Is a summary paragraph placed at the end of the Discussion?

☐   5. Is the significance of the work apparent?

☐   6. Did you compare and contrast your findings with those of other published results?

☐   7. Did you explain any discrepancies, unexpected findings, and limitations?

☐   8. Did you provide generalizations where possible?

☐   9. Did you avoid restating or summarizing the results?

☐   10. Are all elements signaled?

SUMMARY

WRITING A LAB REPORT OR RESEARCH PAPER

1. Understand the "big picture."
2. Obtain instructions and follow them.
3. Collect, organize, and study your references.
4. Make the title succinct, clear, and complete.
5. Follow the IMRAD format.
   - **Abstract:** Include: Question/purpose, experimental approach, results, and conclusion.
   - **Introduction:** Include: Background (unknown for research papers only), question/hypothesis, experimental approach, optional: results and significance.
   - **Materials and Methods:** Provide enough detail to repeat your work.
   - **Results:** Start with your main findings, and then structure the remaining paragraphs. Distinguish between data and results. Report results in past tense.
   - **Discussion:** Start with the key findings and their interpretation; end with a conclusion stating the significance of the work.
6. Signal the elements of the Abstract, Introduction, Results, and Discussion.
7. Provide literature references as needed.

PROBLEMS

Problem 6-1 Abstract

In the following Abstract, identify all essential components and their signals if provided:

To remove cellular metabolic wastes from the body, mammals have an excretory system. To gain more insight into such a system, we dissected a single kidney of a fetal pig. For this purpose, we cut from the lateral margin, leaving the ureter intact and connected to the renal pelvis. We found that the kidney is made up of three different regions internally: the outer cortex, the middle medulla (with the renal pyramids), and the innermost renal pelvis. Blood enters the kidney via the renal artery and is filtered by a maze of small tubules (nephrons) to remove water, ions, nitrogenous wastes, and other materials from the blood. The resulting urine then passed through the collecting ducts into the renal pelvis, which drains into the ureter and ultimately the urinary bladder for excretion. Filtered blood exits the kidney via the renal vein. This study of this system shows how the mammal eliminates its fluid waste through urination.

Problem 6-2 Introduction

Why does this introduction seem incomplete? Identify the known, unknown, question/purpose, and experimental approach. Are these elements clearly identifiable? Why or why not?

Methane clathrate is a solid form of water that contains a large amount of methane within its crystal structure (a clathrate hydrate). Significant deposits of methane clathrate have been found under sediments on the ocean floors[1]. Methane hydrates are believed to form by migration of gas from depth along geological faults, followed by precipitation, or crystallization, on contact of the rising gas stream with cold sea water.

A "Bottom Simulating Reflector" (BSR) was used to detect the presence of methane clathrates along the ocean floor of the Blake Bahama Outer Ridge. Through seismic reflection at the sediment to clathrate stability zone interface caused by the unequal densities of normal sediments and those laced with clathrates, we were able to identify several deposits of methane clathrates at depths of 500 to 1000 m.

Problem 6-3 Results

Assess and revise the following partial Result section. Ensure that all the parts of a paragraph for the Results section are provided and any unnecessary parts are omitted.

To evaluate inhibitory effects of the selected molecules, 10 mM stock solutions of each molecule were prepared in DMSO. A reaction mixture (200 μl) was prepared with the same formula optimized for the enzyme activity assay (0.1 M Tris-HCl pH 8, 0.1 M KCl, 25 mM NaCl, 0.25 mM ATP, and two units of inorganic yeast pyrophosphatase) with 10 μM of the sample molecule. The reaction mixture was incubated for 20 minutes at ambient temperature. Enzymatic reaction was triggered by addition of the substrate B (0.2 mM) and the absorbance of the product was monitored at 290 nm for 10 min.

Six out of 15 sample molecules showed appreciable inhibition at 10 μM (Figure 5). Three of the molecules, A3, A6, and A7, exhibited more than 50% inhibition of the enzyme activity and were further diluted to find the minimal inhibitory concentration. Molecules A3 and A6 exhibited 30% and 45% inhibition of the enzyme activity, respectively, even at 1 μM. DMSO was found not to interfere with the enzyme.

Problem 6-4 Results

Assess the following partial Result section.

1. Identify

   - the purpose or background of the experiment
   - the experimental approach
   - the results
   - the interpretation of the results

2. Are all the parts of a paragraph for the Results section provided? Please explain.

We found that the H384A mutant reduced the k_{cat} value more than 3-fold. The apparent K_m values were increased 7-fold for Fru 6-P and 3.5-fold for PPi. The increase of the K_m values and the reduction of the k_{cat} value of the H384A mutant suggest that the imidazole group of His384 is important for the binding stability as well as for catalytic efficiency of Fru 6-P and PPi substrates.

Problem 6-5 Discussion

Consider the two different opening paragraphs of a discussion about a preventive measure against malaria. Which one is a better first paragraph for a Discussion and why?

Version A:
We trapped and counted the number of mosquitoes within the urban environment of the city of Caracas using conventional carbon dioxide traps. Nearly 70% higher numbers of adult *A. aegypti* were caught in settlements in the vicinity of irrigated urban agricultural sites compared to control areas without irrigated urban agriculture. When we evaluated malaria episode reports from people living in various parts of the city, we found that 18% of malaria cases were reported by people living in the vicinity of urban agricultural areas in the rainy as well as dry seasons, whereas only 2% of the control groups reported incidences of malaria per year.

Version B:
The results of this study show that open-space irrigated vegetable fields in Caracas can provide suitable breeding sites for *Aedes aegypti*. This is reflected in higher numbers of adult *A. aegypti* in settlements in the vicinity of irrigated urban agricultural sites compared to control areas without irrigated urban agriculture. In addition, people living in the vicinity of urban agricultural areas reported more malaria episodes than the control group in the rainy as well as dry seasons. Apparently, the informal irrigation sites of

the urban agricultural locations create rural spots within the city of Caracas in terms of potential mosquito breeding sites.

Problem 6-6 Discussion

Consider the two different concluding paragraphs of a discussion about desert frogs. Which one is a better conclusion for a Discussion and why?

Version A:
In conclusion, this study shows that desert frogs can avoid death by desiccation by maintaining a high body water content and water storage in their urinary bladder and by rapid hydration when water is available. These measures may be employed in combination with behavioral adaptations such as burrowing and change in pigmentation to minimize stresses tending to dehydrate the animals.

Version B:
A limitation of this study was the small number of animals, a single species of frogs, and the location of the study area, which took place in only one oasis in the Mohave Desert. Future studies should be extended to other species, a larger number of animals, and a greater diversity of locations.

CHAPTER 7

Revising and Editing

7.1 GENERAL

Revision is the **key** to successful writing. Generally, the more important the document, the more it should be revised. Once you have written down everything you could think of for a first draft, let it "incubate" for a while. Then it is time to read over it again and revise as needed. The longer you have let it incubate, the more likely you will see passages that you may want to change, recognize portions that need work, think of points to include, and notice those to omit or condense. Good writing takes time, patience, and a lot of hard work.

7.2 REVISING THE FIRST DRAFT

GUIDELINE:

Check the first draft for content and content location first.

When you revise your first draft, check it first for content and organization. Make sure that all the essential points have been included. Everything you say should contribute in some way to the overall purpose and findings (see also Chapter 6), and no steps should have been left out. Any irrelevant points need to be removed, and any missing evidence should be included.

In revising, you essentially work your way from the larger structures of the paper down to the smaller structural elements. Therefore, check the content and organization of the individual sections of the paper first (Introduction, Materials and Methods, Results, Discussion), then the paragraphs, and then sentences. All the parts, paragraphs, and sentences must be in the right order before you revise the style.

The overall structure of your paper should conform to the following outline:

Title:	3–4 important key terms	
Abstract:	**Content: Question/purpose, experimental approach, Results, Interpretation/answer, significance**	
Introduction:	Organization: funnel shape (from broad to specific background information of what is **known**, what remains **unknown**, the **question** you are addressing, the **experimental approach**) **First paragraphs: Background** **Second to last paragraph: Unknown**	
	Last paragraph:	**Question/purpose and experimental approach. Optional are main results and significance.**
Materials and Methods:		Organize chronologically, most to least important, or by subsections
Results:	**1. Paragraph(s):**	**Overview of most important/ interesting result(s)**
	Middle paragraphs:	Describe other results. Organize chronological or most to least important—every result segment should contain purpose of experiment, experimental approach, results, and their interpretation
	Last paragraph:	**State interesting result(s) or summarize main findings if Results section is lengthy**
Discussion:	Organization: Pyramid shape (from specific to more general; interpretation of findings, compare and contrast, models, conclusion, significance)	
	1. Paragraph	**Interpret most important results/ answer to the question of the paper; support and defend interpretation**
	Middle paragraphs:	Chain of topics, compare and contrast findings, list limitations, etc.
	Last Paragraph:	**Conclusion: summarize main findings and significance (future directions)**

Note that in this overall outline, key power positions are written in bold, because they indicate the most crucial structural locations of a paper or report. Pay particular attention to the content and location of these power positions throughout your revisions.

Use the checklist provided for each section of a paper at the end of Chapters 6 to double check that you have included all relevant components in each section. It may help to mark and label important components of each section on a print version of the manuscript to check their completeness. Identifying each essential component will make it apparent to you if there is anything missing or not clearly signaled. Above all, the purpose of your study and the interpretation of your results have to make sense together.

In addition to the overall order listed above, you should consider a secondary order of your topics within each section of your paper. For example, if you are addressing multiple questions in a particular order, keep that order throughout the paper. List your analyses, results, and discussion components in the same order that you introduced your questions. Such order makes it easier for the readers to follow your thoughts.

GUIDELINE:

Check logical organization and flow of sections and subsections.

When you are happy with the structural organization and content of power positions, revise each section for logical organization and flow. Ensure that headings refer to the text they describe. Look at how the ideas are distributed among the paragraphs, and make sure that your arguments are logical. Is it clear how and why the evidence presented supports the interpretation of your findings? Is it clear why a particular experimental approach or technique is appropriate? Have the main concepts been clearly and logically connected?

To check for logical flow, verify that you have a chain of topic sentences running throughout the paper. When read by themselves, the topic sentences should be sufficient to provide a rough outline of the paper.

It may also help to make a reverse outline of your manuscript by going through it paragraph by paragraph. Check that this reverse outline is logically organized.

GUIDELINES:

Revise for style only after you are satisfied with the content and organization.

Pay particular attention to key terms and transitions.

Condense where possible.

Proofread your manuscript.

Once you are satisfied with the content and organization of the first draft, revise it stylistically. You will probably see a lot to change. Here, too, start by working your way from the bigger structures toward the smaller ones.

For good flow between sections and between paragraphs, ensure that the transitions between paragraphs and sections are smooth and that they tie the pieces together. Pay particular attention to key terms and transitions within paragraphs. Add transition phrases and clauses to create the overview of the story (see also Chapter 4 Section 4.2). Then, use the basic writing principles discussed in this book to check for paragraph structure, sentence structure, and word choice. Inch through your manuscript sentence by sentence, word by word. Consider word location (see Chapter 4 Section 4.2). Check whether you have paid attention to either jumping word location or to a consistent point of view.

Look for all possible ways to condense your paper: Omit needless details, redundant words, vague words, and unnecessary paragraphs. If the same concept can be communicated using fewer words, then edit it to the shorter version (see also Chapter 4 Section 4.3). Most readers, editors, and reviewers prefer short, meaty, clear papers. Ensure that you have not repeated any information unnecessarily. Your writing will be more concise if you learn to recognize such repetitions. Finally, proofread the text for punctuation, spelling, and typographical errors.

You will not be able to do all this revising on one draft, so revise in stages. Do as much as you can on the first revision. When you no longer see anything to change, put the paper in a drawer again for a few days. Then you are ready to work on the second draft.

7.3   SUBSEQUENT DRAFTS

GUIDELINE:

Let some time elapse between revisions. Then check for content, logical organization, and style again, and revise if needed.

After you have waited a few days, you will be ready to look at your document with fresh, critical eyes. Start anew by rechecking your draft for content and logical organization, and then recheck for style, especially word location.

Be prepared for additional revisions, particularly if you catch yourself having to reread certain passages. Stumbling across sentences is usually a good indication that more revision may be needed. In revising these questionable passages, check for poor sentence location, sentence structure, word location, and word choice as well as for noun clusters, unclear comparisons, lack of parallel form, change in key terms, lack of transitions, and use of nominalizations. You may also want to read your text aloud to help catch mistakes or awkward phrasing that may otherwise be missed during silent reading.

Revising large sections in a setting will make your document smoother. You may not be able to do this in the first few revisions, but the more you revise, the more you will be able to read through the document in one go.

In addition to revising and proofing your text, pay special attention to your literature cited/references section. Check to see that authors' names are spelled correctly, that authors' initials and citation page numbers are correct, and that references are accurate.

Be aware that the process of revision can be endless. There will inevitably be always something that you would like to change, but you should not spend forever writing one paper. At some point, you have to stop revising. Keep in mind that the writing does not need to be perfect, just clear.

GUIDELINE:

Ensure that you are submitting a complete and final version on time.

When you have finished revising your document, make sure that you have included the final versions of the tables and figures. Recheck that you have followed the Instructions exactly. Pay attention to detail such as font (Times Roman size 10 or 12 is the most preferred), margins (usually 1 in. all around), line spacing (usually double spaced), word count, page numbering, and line numbering if needed.

If there are no more corrections, make sure that the version you submit is really the final version and that it is complete. Ensure also that you turn in your report or article on time.

7.4   CHECKLIST FOR REVISING

Individual sections

Title:
- [ ] Is the title strong?
- [ ] Does it contain 3–4 important key terms?

Abstract:
- [ ] Does the abstract adequately summarize the paper?
- [ ] Have all necessary elements been included (question, experimental approach, results, interpretation/answer, significance)
- [ ] Is the abstract concise?

Introduction:
- [ ] Does the introduction follow a funnel structure?
- [ ] Does the introduction clearly state the overall question of the paper?
- [ ] Does the question follow the unknown?
- [ ] Has the question and experimental approach been stated in the last paragraph?
- [ ] Are all elements (known, unknown, question, experimental approach) clearly signaled?

☐ Did you ensure that the topic has not been reviewed?

Materials and Methods:

☐ Have all experiments been described adequately?

☐ Are experiments organized logically?

Results:

☐ Is the main finding presented in the first paragraph?

☐ Are there errors in factual information, logic, analysis, statistics, or mathematics?

☐ Are all figures and tables explained sufficiently?

Discussion:

☐ Is the overall interpretation of the results clearly stated in the first paragraph?

☐ Did the writer adequately summarize and discuss the topic?

☐ Has a clear conclusion been provided?

☐ Is the significance clearly stated in the last/concluding paragraph?

☐ Is the discussion ordered in a way that is logical, clear, and easy to follow?

References:

☐ Have references been cited where needed?

☐ Are sources cited adequately, appropriately, and accurately?

☐ Are all the citations in the text listed in the References section?

Style and composition

☐ Are the transitions between sections and paragraphs logical?

☐ Are key words repeated exactly?

☐ Are the paragraphs and sentences cohesive?

☐ Has word location been considered?

☐ Did you check for grammar, punctuation, or spelling problems?

☐ Is the style concise?

7.5   EDITING SOMEONE ELSE'S MANUSCRIPT

GUIDELINES:

Provide comments in writing.

Always treat the author with respect.

Editing and evaluating the work of others is one of the best ways to reinforce familiarity with revising strategies. Such editing can also give the author a deeper understanding of how writing affects different readers and how manuscripts are edited professionally. In addition, such editing sets the stage for later peer review as professional scientists.

How you edit a peer's work depends on when in the writing process you are doing the editing. Early drafts should be evaluated primarily with respect to major components of the paper such as the research purpose, the main findings/answer, and the logical overall organization of the paper. Subsequent drafts should be edited for style and composition as well as for flow.

The least helpful comment to receive from a peer revising your work is "It looks OK to me." To be an effective editor, you need to be as specific as possible and point out both strengths and weaknesses. Point out particular places in the paper where revision will be helpful. Do not hesitate to note when something is unclear to you, scientifically or in terms of the writing. If you disagree with the comments of another person who has edited the manuscript, say so. Not all readers react the same way, and divergent points of view can help writers see options for revising.

Always treat the author with respect. Avoid snippy comments such as "So what?" Instead, make suggestions and recommendations on how to improve and strengthen certain passages or on what else to add or omit from the document. If a passage reads well, point out this strength. If an argument is difficult to follow logically or does not make sense, raise objections politely or ask for explanations to clarify the argument. Write your comments either between the text lines or on the margins of the draft. Better yet, use the "Track Changes" option of MS Word because this will clearly show the author where and how to revise a document.

7.6   CHECKLIST FOR EDITING SOMEONE ELSE'S MANUSCRIPT

Checklist for Editing Someone Else's Manuscript

CONTENT

Purpose and Interpretation
- ☐ Is the overall purpose of the paper and/or central question clear?
- ☐ Does the interpretation of the findings answer the overall question of the paper?

Support
- ☐ Is there sufficient evidence to support the answer?
- ☐ Is every paragraph and sentence in the paper relevant to the overall question?
- ☐ Are there portions of the text that could be omitted?

_____

Overall
- ☐ Does the paper advance the field?
- ☐ Does it provide interesting and important insights into the topic of interest?

☐   Have power positions been considered (especially in the Introduction, Results, and Discussion)?

Individual sections

Title:

☐   Is the title strong?

Abstract:

☐   Does the abstract adequately summarize the paper?

☐   Have all necessary elements been included (question, experimental approach, results, conclusion)

☐   Is the abstract concise?

Introduction:

☐   Does the introduction clearly state the overall question of the paper?

☐   Does the question follow the unknown?

☐   Are all elements (known, unknown, question, experimental approach) clearly signaled?

☐   Has the topic been reviewed? _____

Materials and Methods:

☐   Have all experiments been described adequately?

☐   Are methods detailed enough that the study can be repeated by someone else?

Results:

☐   Has the main finding been clearly presented?

☐   Are there errors in factual information, logic, analysis, statistics, or mathematics? _____

☐   Are all figures and tables explained sufficiently?

Discussion:

☐   Has the overall interpretation of the results been clearly stated?

☐   Did the writer adequately summarize and discuss the topic?

☐   Has a clear conclusion been provided?

ORGANIZATION

Overall organization

☐   Is the overall organization of the paper clear and effective?

☐   Are there unclear portions? _____

☐   Could the clarity be improved by changes in the order of the paper? _____

☐   Does the language seem appropriate for its intended audience?

Individual sections

Introduction:

☐   Does the introduction follow a funnel structure?

Materials and Methods:

☐   Are experiments organized logically?

Results:

☐   Is the main finding presented in the first paragraph?

☐   Are all figures and tables labeled properly?

Discussion:

☐   Is the overall interpretation of the results stated in the first paragraph?

☐   Is the significance stated in the last/concluding paragraph?

☐   Is the discussion ordered in a way that is logical, clear, and easy to follow?

References

☐   Have references been cited where needed?

☐   Are sources cited adequately, appropriately, and accurately?

☐   Are all the citations in the text listed in the References section?

STYLE AND COMPOSITION

☐   Are the transitions between sections and paragraphs logical?

☐   Are key words repeated exactly?

☐   Are the paragraphs and sentences cohesive?

☐   Has word location been considered?

☐   Are there any grammar, punctuation, or spelling problems? _____

☐   Is the style concise?

☐   Are there any wordy passages? _____

☐   What other problems exist? _____

OVERALL QUALITY

What are the paper's main strengths? _____

What are the paper's main weaknesses? _____

What specific recommendations can you make concerning the revision of this paper? _____

7.7   SUBMISSION AND THE REVIEW PROCESS

GUIDELINE:

Submit to only one journal at a time.

Upper-level undergraduate students often are in the position of being able to publish their work, either in an undergraduate journal or in an academic journal as part of actual research they may have done in a laboratory. If you are submitting a paper for publication, follow the *Instructions for Authors* provided by your target journal to format your manuscript. Send your manuscript to only one journal at a time. Your paper will only be considered for publication if not submitted elsewhere concurrently. Electronic submissions are standard these days. Instructions for electronic submissions differ from journal to journal—follow them carefully.

The biggest problem in submitting papers electronically is getting figures into the correct format and resolution. Here, it is especially crucial to follow guidelines and suggestions explicitly from the start so that figures and tables do not have to be remade at the last minute to fit the journal specifications. Usually, when you submit figures, you will receive electronic notification that the figure is in the correct format and has been accepted for submission.

Most journals will send an acknowledgment that the paper has been received. The *Instructions to Authors* or the acknowledgment you receive may say how long the journal will take to tell you the fate of the paper. The length of the review process depends very much on the journal. Expect a minimum of two weeks and up to six months to hear back with a decision on whether your article has been accepted or not. Do not submit the paper to any other journal until you get a letter of rejection from your target journal.

When the managing editor has received your manuscript, the editor will assign it to two to three anonymous, qualified reviewers. These reviewers will get back to the editor with specific comments and recommendations on the article. The editor then reads your paper and the comments of the reviewers. The editor may also comment on your paper, summarize the major comments of the reviewers, and state which comments should be taken most seriously and may have to be revised. Most important, the editor decides whether to accept, accept pending revision, reject with encouragement to resubmit, or outright reject your manuscript.

GUIDELINE:

When your manuscript has been accepted, celebrate!

When your manuscript is accepted, the text is set into type by the publisher. This process will take several weeks or months. In the meantime, you may be asked to assign copyright to the journal and to order any reprints.

It is likely that you will not hear from the journal for several weeks or months thereafter. Then the first proofs or galley proofs are sent out to the corresponding author. These proofs are a sample printing of your paper and will look very much the way the paper will appear in the journal. Examine the proofs or their electronic form closely and carefully. Look for errors and make corrections and last minute edits. At this point, you should not introduce any new material nor make any substantial changes. Also, do not make minor changes of wording or emphasis. Usually, only factual errors such as incorrect data in a table should be corrected now. If you need to make substantial changes, discuss this with the editor first. Return any future proofs or an electronic version thereof by the requested date as quickly as possible, usually within 48 hours. The next time you will see your paper it will be in print.

SUMMARY

FIRST REVISION:

1. Check the first draft for content and content location.
2. Check logical organization of sections and subsections. Use checklists at the end of Chapters 6.
3. Revise for style only after you are satisfied with the content and organization.
4. Pay particular attention to key terms and transitions.
5. Condense where possible.
6. Proofread your manuscript.

SUBSEQUENT REVISIONS:

1. Let some time elapse between revisions.
2. Recheck the first draft for content and logical organization.
3. Recheck for style, and revise if needed.
4. Ask for comments and constructive criticism in writing.
5. Ensure that you are submitting a complete and final version on time.

REVIEWING A MANUSCRIPT:

1. Provide comments in writing.
2. Always treat the author with respect.

SUBMISSION:

1. Submit to only one journal at a time.
2. When your manuscript has been accepted, celebrate!

CHAPTER 8

Sample Reports

8.1 GENERAL

This chapter provides two well-written, sample reports. The first exemplifies a typical lab report written during a weekly undergraduate laboratory course in biochemistry. The second has been composed as a final paper for a longer-term inquiry-based project spanning about two-thirds of a semester-long microbiology lab course. Note that these are sample reports, and different schools and instructors may have somewhat different requirements on what to include or display.

Both reports follow the overall layout of typical scientific research articles, the IMRAD format (see Chapter 6), but the format of the weekly lab report is less stringent and also contains raw data and calculation steps. The final microbiology paper has almost all characteristics of a well-written scientific research paper in terms of format as well as style and tone of the paper. Annotations on the margins of each of these reports point out important passages and elements to be considered in writing your own reports.

8.2 SAMPLE WEEKLY LAB REPORT

Example 8-1

A clear title as well as your name and course number should identify the lab report.

Tammy Wu

Bio 106L, 2008, week 6

LABORATORY 6: TIME COURSE OF β-GALACTOSIDASE INDUCTION

The Abstract provides a general overview of the experiment and the findings.

Abstract

In this experiment we measured the induction of the lac operon of *E. coli* when induced with IPTG (isopropyl thiogalactoside). When IPTG is added to the lac operon, the repressor–operator complex binding to the operon is dissociated and mRNA synthesis begins immediately. By adding the substrate ONPG (orthonitrophenyl galactoside), which is cleaved by β-galactosidase to generate the yellow compound *o*-nitrophenol, we can measure β-galactosidase activity by determining A_{420} and A_{550} absorbance. Thus, the time course of β-galactosidase synthesis can be revealed.

Introduction

β-galactosidase is an enzyme that catalyzes the hydrolysis of β-galactosides into monosaccharides. In *E. coli*, β-galactosidase is encoded by the *lacZ* gene, which is part of the *lac* operon. The operon is activated in the presence of lactose when glucose levels are low. Lac Z codes for the β-galactosidase monomer; active β-galactosidase is a tetrameric protein.

The Introduction of lab reports is usually much shorter than that of research articles written for publication. This introduction provides a general background. It also states the purpose of the experiment as well as the general experimental.

β-galactosidase is frequently used in genetics, molecular biology, and other life sciences as a reporter marker in various applications. In the β-galactosidase assay an active enzyme may be detected using the substrate X-gal, which forms an intense blue product after cleavage by β-galactosidase. β-galactosidase can also cleave ONPG (orthonitrophenyl galactoside), leading to the yellow product *o*-nitrophenol. Production of the enzyme can be induced by IPTG, which binds and releases the lac repressor from the lac operator. This release allows transcription to proceed.

In this experiment, we determined the time course of β-galactosidase synthesis by measuring ONPG cleavage during different time points after IPTG induction.

Materials and Methods

IPTG Induction

In the Materials and Methods section, experimental setups and execution are described; the purpose for each step is also included.

To determine β-galactosidase enzyme activity for our experiment, we induced *E. coli* cells with IPTG. IPTG induction was stopped at various times by adding Z medium (containing β-mercaptoethanol; stops transcription) and toluene (allows ONPG to permeate cells). ONPG was then added and A_{420} and A_{550} measured to assay enzyme activity. Background absorbance due to cell density was obtained by determining A_{600} of the grown *E. coli* culture twice. Background β-galactosidase activity (tube 2) was assessed by assaying *E. coli* cells to which ONPG had been added but without IPTG induction.

Calculation of Raw Enzyme Units

Subheadings are often used in the Materials and Methods section

To normalize for cell density at each time point, each A_{600} value was determined by graphing A_{600} before and after induction versus time, generating a standard curve (Figure 1 [Figure 8.1], Table 1 [Table 8.1]).

Raw enzyme units were calculated as follows:

Formulas and
calculation steps
are laid out

$$\text{Units enzyme} = 1000 \times (A_{420} - 1.75 * A_{550})/(t \times v \times A_{600})$$

where $t = 54$ min for tubes 1–8, $t = 14$ min for tubes 9–13, and $v = 0.5$ ml.

Determination of Transcription and Translation Rates

Background β-galactosidase enzyme units (tube 2) were then subtracted from each raw enzyme unit value, giving enzyme units from IPTG induction alone (Figure 2 [Figure 8.2]). By extrapolating this rate, transcription and translation rates were found (Table 2 [Table 8.2], Table 3a [Table 8.3a], Table 3b [Table 8.3b]).

Measurements and Data

Table 1 [Table 8.1] Cell Density Control Values

TIME	$A_{(600)}$
-7	0.38
62	0.29

Unlike scientific
research papers
destined for publi-
cation, lab reports
usually contain
original raw data
and calculations to
let your instructor
evaluate how and
if you arrived at
your conclusions
correctly. Note that
for publication,
grid lines should
not be used.

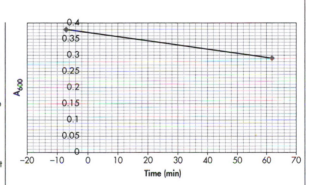

Figure 1 [Figure 8.1] Background absorbance due to cell density.

Table 2 [Table 8.2] Extrapolated $A_{(600)}$ Values

TUBE #	$A_{(600)}$
1	—
2	0.375
3	0.37
4	0.365
5	0.362
6	0.36
7	0.352
8	0.345

TUBE #	$A_{(600)}$
9	0.335
10	0.32
11	0.31
12	0.295

Table 3a [Table 8.3a] Calculated Enzyme Units

TUBE #	TIME AFTER INDUCTION (MIN)	ENZYME UNITS
1		
2	-1	7.28
3	1	4.33
4	4	8.95
5	6	10.36
6	10	5.38
7	15	17.81
8	20	10.31
9	30	28.89
10	40	27.79
11	50	27.3
12	60	36.08
13	70	27.0

Table 3b [Table 8.3b] Corrected Enzyme Units = Total Enzyme Units – Enzyme Units from Tube 2 (Background)

TUBE #	TIME AFTER INDUCTION (MIN)	CORRECTED ENZYME UNITS
1		
2		0
3	1	
4	4	1.67
5	6	3.08
6	10	
7	15	10.53
8	20	3.03
9	30	21.61
10	40	20.51
11	50	20.02
12	60	28.8
13	70	20

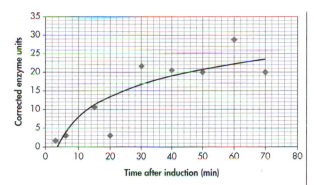

Figure 2 [Figure 8.2] β–galactosidase activity after IPTG induction of wild-type *E. coli* cells.

Results and Discussion

In our experiment, according to the graph of β-galactosidase activity versus time of induction (Figure 8.2), enzyme activity was first noted at approximately 2 min after induction. Approximately 30 min after induction with IPTG, β-galactosidase activity started to level off and reached a saturation point. After 30 min, the assays all gave enzyme units over 20 but did not show the same increase rate as the previous times. This is likely due to the enzyme operating at maximum capacity, but could also be caused by the repressor binding to the operon and dissociation of cAMP, cessation of translation, or decreased viability of the enzyme (e.g., tetramer dissociation). All mechanisms could be mediated by the cell's own regulatory mechanisms for β-galactosidase activity.

Our experiment had some limitations. To graph a better curve, more samples at more time points would need to be taken, and experiments should be repeated. Aberrant values observed in our measurements and graph could have been due to nonhomogeneity of *E. coli* cells, or errors in adding reagents or in reading absorbance values.

Because multiple copies of the lac mRNA exist, and would be translated simultaneously, tetramer assembly should be much faster than translation of the lacZ gene into the β-galactosidase monomer. The rate-limiting step in the start of enzyme activity, which took place about 2 min (120 s) after IPTG induction, was therefore assumed to be the protein synthesis of the enzyme monomer. Thus, the minimal rate of chain elongation of protein synthesis would be 1170 amino acids (aa) in the monomer divided by 120 s, or 9.75 aa/s. Assuming that transcription and translation are coupled, the minimal rate of mRNA synthesis would be 1170 x 3 / 120 = 29.25 bases/s.

A combined Results and Discussion section lets you state and interpret your findings and discuss potential limitations; often it also lets you answer specific questions that may have been posed by your instructor.

Different from professional scientific research papers or term papers, your lab report usually does not compare and contrast your findings and their meaning with those of others in the field.

Because the (Results and) Discussion section of a lab report are rather short and focused on your single experiment only, this section often does not contain a concluding final paragraph.

If an adenyl cyclase mutant instead of wild-type *E. coli* cells would have been used, induction of the lac operon would not be seen because adenyl cyclase produces cAMP. The dissociation of the repressor–operon complex is cAMP-dependent; there is a CRP-cAMP binding site upstream of the RNAP and repressor sites, and binding of cAMP leads to transcription of the gene.

8.3   SAMPLE PROJECT PAPER

Example 8-2

Author's name	Amanda Miller
Course	Microbiology
Date	4/28/11

Informative and complete title

Isolation and Characterization of *Bacillus amyloliquefaciens* from Facial Skin

Abstract

Abstract is less than 250 words, and it contains all essential elements and no citations.

Background

Question/purpose

Experimental approach

Results

Conclusion

Microflora are involved in a mutually beneficial relationship with the human body, where the microflora thrives in an ideal environment and the body gains protection against pathogens and receives aid in several processes. The purpose of this experiment was to identify one of the bacteria that inhabit the skin of the human body, specifically the facial skin, and to determine whether it is pathogenic, normal microflora, or a harmless transient bacteria. A sample of bacteria was taken from the cheeks, isolated, and analyzed with respect to morphology, Gram status, metabolic capabilities, antibiotic sensitivity, growth curve, and species identification with a Biolog assay. Biolog analysis revealed that the identity of the isolate was *Bacillus amyloliquefaciens*, transient soil bacteria associated with plants. Characterization of the isolate was highly consistent with the characterization of *B. amyloliquefaciens*, with the exception of some minor metabolic capabilities and morphology, indicating accurate species identification.

Introduction

First paragraph provides background information, which is paraphrased and cites sources.

As soon as a human being is born, he or she comes into contact with microorganisms that colonize the body (Chiller et al., 2001). The human body and these microbes enter into a symbiotic relationship from which both gain some benefit. These microbes, which thrive and do not generally harm the host if colonized in the correct area or environment on the body, are known as normal human microflora (Willey et al., 2008).

The relationship between the microflora and the human body is beneficial to the microflora in that it provides an environment for them to grow and gain nutrients. This relationship is beneficial to the human host because the bacteria provide a source of protection against harmful biological agents, such as pathogenic bacteria, and aid in normal bodily processes. This relationship allows the correct balance of microflora to thrive on and inside the human body without causing it harm and instead increasing its health.

The bacteria that inhabit the body are not uniformly distributed; some areas have more species of microflora than others, and not all microflora can exist on all parts of the body. The areas that they colonize on the human body are, usually, the skin and mucus membranes of the oropharangeal tract, the gastrointestinal tract, and the female genital tract (Sullivan et al., 2001). The ideal environment for growth and survival differs among species, which is why each area of the body that supports normal growth has a different set of microflora (Tancrede, 1992). For example, the skin is generally dry and slightly cooler than the inside of the body, and has a slightly acidic pH (Chiller et al., 2001). The skin also has high concentrations of NaCl and can secrete inhibitory substances to prevent some bacterial growth (Willey et al., 2008). Just on the skin itself, varying conditions exist, especially in terms of moisture, pH, exposure, and capability to trap dirt (Evans et al., 1950). For example, the hands or fingernails may be more exposed to bacteria than skin on the face.

> Sources cited in parentheses, name followed by year

The microbes able to survive and benefit from using the skin as a host, therefore, differ from areas of the body which provide different environmental conditions. For example, the moist, often mechanically disrupted, neutral pH environment that is provided by the mouth will host different bacterial numbers and species than the skin, which is dry and acidic. Normal skin bacteria also differ from those in the GI tract. The microbes that may inhabit the large intestine are usually anaerobes, as they are not exposed to the external environment, while skin microbes are often found on superficial cells exposed to the air and therefore are frequently aerobic. Microbes on the skin and in the intestines are similar in that they must be continuously replaced, but for different mechanical reasons. Intestinal microbes are replaced as material moves through the intestine and transports them out. On the skin, microbes are replaced because the dead skin cells they adhere to are discarded frequently and are replaced themselves (Willey et al., 2008). Essentially, variances in the environmental conditions provided by the human body promote diversity in the microflora that uses it as a host.

> Note italics for genus and species names; capitalizing genus but lower case species name

Several types of microbes are generally found at each site on the body. On the skin, Gram-positive bacteria predominate, including, but not limited to, *Staphylococcus aureus*, *Bacillus* species, *Streptococcus*, coagulase-negative bacteria like *S. epidermis* and *S. hominis*, *Micrococcus* species such as *M. luteus*, *Corynebacterium*, and *Propionibacterium*

(Chiller et al., 2001). Although they are not the most common, the latter two are involved in causing acne lesions on the skin (Marples, 1974).

It is important to determine the distribution of microflora on the human body because of their role in protecting the body from pathogens and infection and in aiding internal processes such as digestion. Without the proper balance of microbes, the body will not be able to defend itself effectively and will not be able to break down materials needed for survival (Sullivan et al., 2001). This knowledge is especially important when considering the use of antibiotics. If the normal microflora is susceptible to the type of antibiotic being used to treat an infection, there is a chance that the normal microflora will be killed or significantly reduced (Edlund and Nord, 2000). Elimination of normal microflora diminishes its ability to protect the body by competing for nutrients, blocking the adherence of pathogens, producing toxins and inhibitory chemicals, and promoting the body's natural ability to produce antibodies and other defense mechanisms against pathogens (Chiller et al., 2001). The resulting lack of colonization resistance to opportunistic microbes such as *Clostridium difficile*, which can cause infection and disruption of the intestinal tract, can be extremely harmful (Sullivan et al., 2001). It is also important to study the distribution of microflora in order to understand the detrimental effects of a usually normal microbe growing in the wrong place. For example, if a bacterium that normally exists only on the skin gets into the bloodstream or a tissue that is usually sterile, that bacterium becomes a potentially harmful opportunistic pathogen (Willey et al., 2008)

The purpose of this experiment was to isolate a single strain of bacteria from one site on the body—the skin on the outside of the cheeks, which is known to carry normal microflora—and to discover the identity of the isolate. To determine the strain of bacteria isolated at this site, several procedures and tests were performed, including phenotypic characterization (e.g., Gram status), analysis of metabolic capabilities, antibiotic sensitivity, growth curve, and ability to use varying carbon sources of chemical sensitivity assays in a Microlog plate.

Materials and Methods

Isolation

A sterile swab was dipped in sterile dH$_2$O to moisten it before it was rubbed on the skin located on the outside of the cheek. Using sterile technique, a nutrient agar plate was inoculated by rubbing the sample swab across the surface of the plate, which was then incubated for 43 h at 37°C. Colony types were recorded, and one type was chosen for further isolation. A second nutrient agar plate was streaked for single colonies of the selected type and then incubated at 37°C for 24 h. A third nutrient agar plate was streaked again for single colonies of the selected type and was then incubated again at 37°C for 29 h. A fourth

Unknown/problem/need is stated.

In the last paragraph, the purpose of the experiment is clearly signaled and stated.

Experimental approach is stated.

This section is further divided into subsections, each with its own specific heading.

The section provides sufficient details for other scientists to repeat the experiments; for example, every incubation step specifies temperature and time.

plate was re-streaked for single colonies and then incubated at 37°C for 48 h to ensure that the isolate was pure.

Gram Stain/Morphology Identification

Sterile nutrient broth was inoculated with a single colony of the isolate AJM-3. After 24 h, a Gram stain of both the nutrient broth culture and the plate culture was performed. Using sterile technique, a smear from both cultures was placed within dime-sized circles on a glass slide (the plate sample was combined with a small drop of sterile water on the slide), allowed to dry, and was heat-fixed by passing the slide through the flame of a Bunsen burner. *S. marcescens* and *B. cereus* were also fixed on the slide as negative and positive controls, respectively. The slide was covered with enough drops of primary stain crystal violet to completely cover the samples and allowed to stain for 1 min. The stain was rinsed with dH$_2$O before the slide was covered again with several drops of Gram's iodine mordant, which was also allowed to stand for 1 min. The mordant was rinsed with dH$_2$O, and ethanol decolorizer was applied by running it across the slide for about 20 s, until most of the purple stain stopped running off of the slide. The slide was immediately rinsed with dH$_2$O and counterstained with safranin, which was rinsed with dH$_2$O after 1 min. The slide was carefully blotted dry with bibulous paper and observed under the 100 x oil objective for cellular morphology and Gram status, compared to the positive and negative controls. The presence of dark purple coloration was considered Gram positive, while pink coloration was distinguished as Gram negative. Colony morphology was identified by observation of the bacteria growing on an agar plate.

Oxidase Test

An oxidase test disk was placed on a clean glass slide using sterilized forceps. The disk was moistened by dipping it in sterile dH$_2$O and placed back on the slide. A moderate amount of the isolate AJM-3 was sampled with a sterile inoculating loop and placed on the disk. This process was also performed on *E. coli* and *B. catarrhalis* as negative and positive controls, respectively. The disks were incubated at room temperature for 5 min and observed. Presence of purple color indicated cytochrome *c* oxidase was produced, while a lack of change in color was considered a negative result.

Control experiments (negative and positive) are included.

Catalase Test

An entire colony of AJM-3 was transferred onto a clean glass slide using an inoculating loop and sterile technique. One drop of 3% H$_2$O$_2$ was added on top of the sample and observed. The same procedure was performed on *E. coli* and *L. lactis* as controls. Presence of bubbles indicated that the bacterial species was positive for catalase production. Lack of bubbling indicated that the species does not produce catalase.

Nitrate Reduction Test

Tubes of nitrate broth were inoculated with AJM-3, *B. subtilis*, *E. coli*, and *L. lactis*, and one tube was left uninoculated as a control. The tubes were incubated at 37°C for 48 h. The cultures were removed from incubation, and 5 drops of Nitrate Reagent A were added, along with 5 drops of Nitrate Reagent B. The tubes were swirled and observed for color changes. A pinch of zinc dust was added with forceps to the tubes that did not change to red initially, and color was recorded again after swirling the tubes and letting them incubate for 10 min at room temperature. The cultures that initially changed color to red were determined to be positive for nitrate reduction. The cultures that changed to red after the addition of zinc were determined to be negative for nitrate reduction. Cultures whose color remained the same were classified as capable of complete denitrification.

> Every experiment described in the Materials and Methods section refers to a relevant experiment reported in the subsequent Results section.

Mannitol Salt Agar Test

AJM-3, *E. coli*, *S. epidermis*, and *S. aureus* were streaked, using sterile technique, onto an MSA plate divided into thirds. The plate was incubated at 37°C for 48 h and then observed for bacterial growth and change in media color. Presence of growth indicated that the species could tolerate the high salt concentration, and a change in the media color from red to yellow indicated that the bacteria could ferment mannitol. Lack of color change indicated that the bacteria were incapable of fermenting mannitol. AJM-3 was compared to the other three strains as controls.

MacConkey Agar Test

AJM-3, *E. coli*, *S. epidermidis*, and *S. marcesens* were streaked, using sterile technique, onto an MAC plate divided into thirds. The plate was incubated at 37°C for 48 h and observed for bacterial growth and color of bacterial growth. Presence of growth indicated that the bacteria were not inhibited by bile salts and crystal violet. Bacteria that appeared purple or red in color indicated that they were able to ferment lactose. Bacteria that appeared white or yellow indicated that they were unable to ferment lactose. AJM-3 was compared to the other three strains as controls.

> For every experimental approach, potential outcomes and their meaning are described

Starch Hydrolysis Test

An agar plate containing starch and iodine as an indicator reagent was inoculated with AJM-3, and *E. coli* and *B. subtilis* as controls, using sterile technique. The plate was incubated at 37°C for 48 h and observed for bacterial growth. Bacteria were determined to be positive for starch hydrolysis if they showed significant growth and formed a "halo" around the bacteria where the starch was hydrolyzed. Bacteria that showed little to no growth and had no "halo" surrounding the growth were considered negative for starch hydrolysis.

Antibiotic Sensitivity Test

A nutrient broth tube was inoculated with AJM-3 using sterile technique. 45 hours later, a sterile swab was dipped into the broth culture and rolled along the inside of the tube to rid the swab of excess moisture. A Mueller-Hinter agar plate was lightly swabbed in tight streaks and then rotated 1/3 of the way around. This process was repeated twice to create a lawn of bacteria, and the outside edge of the plate was swabbed again to ensure complete coverage. The plate was dried at room temperature. Antibiotic disks of Erythromycin (15 µg), Penicillin (10 units), Rifampin (5 µg), Streptomycin (10 µg), Tetracycline (30 µg), and Vancomycin (30 µg) were placed in a circular pattern on the plate using sterile forceps. The plate was incubated lid up for 48 hours. Zones of inhibition were measured and compared to the standards established by the Clinical and Laboratory Standards Institute to determine whether the bacteria were resistant, intermediate, or susceptible to each antibiotic (Madigan et al. 2009).

Growth Curve and Generation Time

To follow the cell growth for the isolate AJM-3, spectrophotometric determination of the optical density of a culture and the plating of a dilution series to count colony-forming units (CFUs) were used. In the spectrophotometric determination of the growth curve, a single colony of AJM-3 was inoculated into a sterile broth tube and incubated for 36 h at 37°C. 1–5 ml of the overnight culture was added to 50 ml of sterile nutrient broth and incubated again for another 282 min. At this time, the culture was removed from incubation, and 2 ml was aseptically transferred to a clean cuvette. The culture was returned to incubation. A spectrophotometer set to 600 nm was blanked with a cuvette containing 2 ml of sterile nutrient broth, and the optical density of the 2 ml of culture was measured. This process was repeated every 20–30 min for 3–4 hours until optical density was measured at 8 time points after the sterile nutrient broth was added. One optical density measurement was taken at a later time, 670 min after the sterile nutrient broth was added, for a total of 9 time points.

In the CFU count determination of the growth of the cells, 1 ml of the broth culture of AJM-3 was removed 285 min after the sterile nutrient broth was added and transferred to a sterile microfuge tube. The sample was placed on ice and then diluted to 10^{-3}, 10^{-4}, 10^{-5}, 10^{-6}, and 10^{-7} dilutions of the original culture. These dilutions were then plated using the drip method; 10 µl of each dilution was pipetted in a line of drops across the top of the plate, allowed to run down the plate, and stopped before reaching the edge. The plate was allowed to dry, inverted and incubated for 48 h. This procedure was repeated every 40–60 min for 3–4 h until 4 time points were taken following the sterile nutrient broth addition. One CFU time point was taken at a later time, 685 min after the sterile nutrient

---

Although most Materials and Methods sections contain few if any references, citations can and should be added whenever appropriate.

broth was added, for a total of 5 time points. After 48 h, the CFUs were counted in as many drips as possible, and the CFUs/ml were calculated at each time point.

To calculate the number of divisions or generations (n), the values for optical density at the first and last time points were used in the formula $n = (\log N - \log N_0)/\log 2$. This value and the total time (t) between the first and last time points were used in the formula for generation time, $g = t/n$.

Microlog Species Identification

A fresh nutrient agar plate was streaked with AJM-3 using sterile technique and incubated at 37°C for 21 h. A turbidometer was calibrated with an 85% turbidity standard. The turbidometer was blanked (adjusted to 100%) with the tube of IF-A. The IF-A tube was inoculated with AJM-3 by touching the tip of a swab to a single colony of the bacteria from the fresh plate and gently rubbing the swab on the bottom of the tube. The IF-A was recapped and inverted, and any extra particles were broken up with a new sterile swab. IF-A was placed in the turbidometer in the same position as when it was used to blank. Turbidity was measured at 98.5. The inoculum was poured into a sterile reservoir. A multichannel pipette was used to inoculate the MicroPlate with 100 µl of inoculums in each well. The plate was covered with its lid and incubated for 23 h.

The results of the assay were scored, comparing the A1 negative control to columns 1–9 and the A10 positive control to columns 10–12. Scoring was performed manually. For columns 1–9, "++" indicated very dark gray coloration, "+" indicated significantly darker coloration than the control well, "/" indicated coloration ambiguous in comparison to controls (very slightly darker or lighter), "–" indicated coloration the same as control, and "0" indicated a lighter coloration than the control. For columns 10–12, "+" indicates purple coloration similar to the positive control, "–" indicates purple coloration with less than half the color of the positive control, "/" indicates ambiguity in the coloration in comparison to the positive control.

When entering the MicroLog plate data into the Biolog software, "++" and "+" were entered as positive, "0" and "–" were entered as negative, and "/" was entered as intermediate. The species ID, probability, and similarity identified through Biolog were recorded.

Results

The analyses of these discriminating tests were used to determine the identity of a microbe found on the human body. The results of morphology observation, Gram status, metabolic capabilities, antibiotic sensitivity, growth rate analysis, and a Biolog Assay were used to determine the specific species found on the site.

Overview paragraph—not always needed but helpful for long Results section such as this

Isolation

The sample collected from the skin on the outside of the cheek, after incubation, yielded five different colony types. Type 1 was a large, cream-colored singular colony 6 mm in width, with irregular, raised, and undulate morphology. Type 2 included 30 small, pale yellow colonies evenly distributed over the plate, measuring about 0.5 mm in width, with circular, raised, and entire morphology. Type 3 colonies were also about 30 in number, were evenly distributed, and had circular, raised, and entire morphology. However, they were slightly larger than Type 2, measuring 1 mm in diameter, and were cream-colored. Type 4 included 40 colonies that were slightly less than 1 mm and were cream-colored, displaying circular, raised, and entire morphology. Type 5 included about 200 cream-colored colonies much less than 1 mm in diameter, with circular, raised, and entire morphology.

Type 1 was selected for isolation. After streaking for a single colony of Type 1, the plate contained only the Type 1 colony type. Type 1 colony was re-streaked a second and third time and was the only colony type on these plates as well, confirming a successful isolation of this specific bacterial species, temporarily named AJM-3.

Phenotypic Characterization

Morphology, analyzed by Gram stain and streak plate observation, is helpful in finding correlations between the appearances of bacteria that are being compared. After further observation, the isolated bacteria showed irregular, raised, and undulate morphology consistent with previous observations. Gram staining revealed AJM-3 to be Gram-positive rods linked together in chains, sometimes 2-cell or 4-cell long.

Metabolic Characterization

Metabolic activities were measured in order to compare the specialized capabilities of the isolate. A summary of the metabolic activity of AJM-3 can be seen in Table 1 [Table 8.4].

Table 1 [Table 8.4] Catabolic Activities of Unknown Bacteria AJM-3, Tested by the Oxidase Test, Catalase Cest, Nitrate Reduction Test, Mannitol Salt Agar Test, MacConkey Agar Test, and Starch Hydrolysis Test.

TEST	AJM-3
Cytochrome c oxidase production	Positive
Catalase production	Positive
Nitrate reduction	Reduction to NOs
Mannitol fermentation	Bacterial growth, no fermentation
Lactose fermentation	Fermentation, small amount of growth
Starch hydrolysis	Hydrolysis

As seen in Table 1 [Table 8.4], AJM-3 produces both cytochrome *c* oxidase and catalase in its respiration processes. The culture of the broth turned red upon the addition of Nitrate Reagents A and B in the Nitrate Reduction Test, indicating that they reduced nitrate to nitrite, but are not capable of complete denitrification. The metabolic tests also indicated that AJM-3 is able to hydrolyze starch and that the species can ferment lactose but not mannitol. The mannitol test was somewhat ambiguous, as the media colors from the adjacent bacteria samples ran together in some places. The MacConkey Agar Test for lactose fermentation showed only a little bit of bacterial growth, but the color of the bacteria was purple, indicating that it was still able to ferment the lactose.

All findings have a corresponding Material and Methods section.

Results are not only listed but also briefly interpreted for the reader without going into any further discussion.

Antibiotic Sensitivity

Figures and tables are referred to in parentheses.

The Kirby–Bauer test, intended to address the specialized resistance abilities of an isolate, revealed that AJM-3 had different levels of sensitivity for each of the six antibiotics tested (Table 2 [Table 8.5])

Table 2 [Table 8.5] Diameters of Zones of Inhibition and Assessment of Sensitivity of the Unknown Bacteria AJM-3 to Disks of the Antibiotics Erythromycin (15 µg), Penicillin (10 units), Rifampin (5 µg), Streptomycin (10 µg), Tetracycline (30 µg), and Vancomycin (30 µg) on Mueller–Hinton Agar plates

ANTIBIOTIC	AJM-3
Erythromycin	30 mm (susceptible)
Penicillin (Staphylococci)	26 mm (resistant)
Penicillin (Others)	—
Rifampin	19 mm (intermediate)
Streptomycin	21 mm (susceptible)
Tetracycline	17 mm (intermediate)
Vancomycin (Staphylococci)	22 mm (susceptible)
Vancomycin (Enterococci)	—

AJM-3 best resembled *Staphylococci,* based on the results of the Mannitol Agar Test in comparison to the control, and therefore was analyzed under the corresponding range for the zones of inhibition for Penicillin and Vancomycin.

AJM-3 was resistant only to Penicillin but was susceptible to Erythromycin, Streptomycin, and Vancomycin. AJM-3 had intermediate antibiotic resistance to Rifampin and Tetracycline (Table 2 [Table 8.5]).

Growth Curve and Generation Time

CFU counts and optical density measurements at specific time points were used to analyze the growth rate capabilities of the isolate. As seen in Figure 1 [Figure 8.3], approximately 300 min after the overnight culture was added, the bacteria appeared to be in log phase of growth, which continued until optical density was measured at the last time point (about 700 min). The optical density growth curve data were estimated to have a better representation of the growth curve of AJM-3, as it had more data points and a more consistent pattern of growth over time, than the CFU growth curve data. The CFU growth curve data were inconclusive, and therefore generation time was calculated using the OD data.

Data are clearly visible (darkest entity).

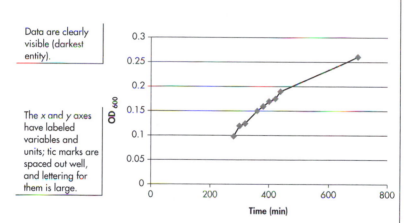

The x and y axes have labeled variables and units; tic marks are spaced out well, and lettering for them is large.

Figure caption is directly below the figure, and it contains a title as well as a legend/descriptive text.

Figure 1 [Figure 8.3] Optical density growth curve
Optical Density measured with a spectrophotometer at a wavelength of 600 nm at intervals after the addition of 5 ml of overnight culture of AJM-3 to 50 ml of sterile nutrient broth. Identified as log phase of the cell cycle.

Generation time for AJM-3, calculated from Figure 1 [Figure 8.3], was 158.9 min.

Species Identification

To comprehensively measure the sensitivity and utilization capabilities of the isolate, Biolog software was used to analyze a Microlog plate for AJM-3. The MicroLog plate used for bacterial identification was manually scored and entered into the Biolog system (Figure 2a [Figure 8.4a] and Figure 2b [Figure 8.4b]). A positive identification was made, indicating that AJM-3 was an isolate of the species *Bacillus amyloliquefaciens*. Probability of a correct identification was 0.957, and similarity was measured at 0.660.

	1	2	3	4	5	6
A	Negative Control faint gray	Dextrin	D-maltose	D-Trehalose	D-Cellobiose	Gentiobiose
		++	++	++	++	++
B	D-Raffinose	D-Lactose	D-Melibiose	Methyl-D-Glucoside	D-Salicin	N-Acetyl-D-Glucosamine
	+	+	/	++	++	++
C	D-Glucose	D-Mannose	D-Fructose	D-Galactose	3-Methyl Glucose	D-Fucose
	++	++	++	/	+	/
D	D-Sorbitol	D-Mannitol	D-Arabitol	myo–Inositol	Glycerol	D-Glucose-6-PO$_4$
	+	+	–	+	+	/
E	Gelatin	Glycyl-L-Proline	L-Alanine	L-Arginine	L-Aspartic Acid	L-Glutamic Acid
	++	++	++	++	++	++
F	Pectin	D-Galacturonic Acid	L-Galactonic Acid Lactone	D-Gluconic Acid	D-Glucuronic Acid	Glucuronamide
	++	++	0	++	+	+
G	p--Hydroxy Phenylacetic Acid	Methyl Pyruvate	D-Lactic Acid Methyl Ester	L-Lactic Acid	Citric Acid	Keto-Glutaric Acid
	–	+	/	+	+	/
H	Tween 40	Amino-Butyric Acid	Hydroxy-Butyric Acid	Hydroxy-D,L Butyric Acid	Keto-Butyric Acid	Acetoacetic Acid
	+	/	0	–	0	+

Chemical names are used; controls are included

Figure 2a [8.4a] Manual scoring of Biolog assay from a GENIII MicroPlate, each well holding 100 μl of IF-A inoculating fluid containing AJM-3. "++", very dark gray coloration; "+", significantly darker coloration than the control well; "/", coloration ambiguous in comparison to controls (very slightly darker or lighter); "-", coloration the same as control; and "0", a lighter coloration than the control.

Figure caption starts with a title followed by additional relevant information such as explanations for what abbrevations stand for

Discussion

First paragraph contains main findings

The species identified by the Biolog system, *Bacillus amyloliquefaciens*, is a bacteria normally found in the soil (Idriss et al., 2002). It is very closely related to *B. subtilis*, and was ultimately differentiated as a separate species not only because of DNA differences, but largely because of its ability to grow in certain environments (for example, in high concentrations of NaCl), its comparatively greater production of the hydrolyzing enzyme α-amylase, which breaks down polysaccharides, and its ability to ferment lactose (Welker and Campbell, 1967).

	7	8	9	10	11	12
A	Sucrose ++	D-Turanose +	Stachyose –	Positive Control light gray	pH 6 +	pH 5 +
B	D-Acetyl-D Man-nosamine ++	N-Acetyl-D Galactos-amine –	N-Acetyl Neuraminic Acid 0	1% NaCl +	4% NaCl +	8% NaCl /
C	L-Fucose /	L-Rhamnose /	Inosine –	1% Sodium Lactate +	Fusidic Acid –	D-Serine –
D	D-Fructose 6-PO4 +	L-Aspartic Acid ++	D-Serine –	Trolean-domycin –	Rifamycin SV –	Minocycline –
E	L-Histidine ++	L-Pyroglutamic Acid +	L-Serine ++	Lincomycin –	Guanidine HCl +	Niaproof 4 –
F	Mucic Acid +	Quinic Acid ++	D-Saccharic Acid +	Vancomycin –	Tetrazolium Violet –	Tetrazolium Blue –
G	D-Malic Acid –	L-Malic Acid ++	Bromo-Succinic Acid ++	Nalidixic Acid –	Lithium Chloride +	Potassium Tellurite +
H	Propionic Acid /	Acetic Acid ++	Formic Acid ++	Aztreonam –	Sodium Butyrate +	Sodium Bromate –

Figure 2b [8.4b] Manual scoring of Biolog assay from a GENIII MicroPlate, each well holding 100 μl of IF-A inoculating fluid containing AJM-3. Scoring for columns 7–9: "++" very dark gray coloration, "+", significantly darker coloration than the negative control well; "/", coloration ambiguous in comparison to the negative control (very slightly darker or lighter); "–", coloration the same as the negative control; and "0", a lighter coloration than the negative control. Scoring for columns 10–12: "+", gray coloration similar to the positive control; "-", gray coloration with less than half the color of the positive control; "/", ambiguity in the coloration in comparison to the positive control.

B. amyloliquefaciens is normally associated with plants, and often serves as a biocontrol agent for the plants it colonizes. The bacteria colonize many types of plants and are especially important for the agricultural industry, as they protect the crops from disease and increase growth success (Arguelles-Arias, 2009). These bacteria produce antimicrobial dipeptides or cyclic lipopeptides (Yin et al., 2011). Such antimicrobial agents protect some plants against phytopathogens such as fungi, yeasts, and other bacteria (Yin et al., 2011). The capability of *B. amyloliquefaciens* to promote plant growth and protect from pathogens indicates its potential for involvement in developing biocontrol agents (Arguelles-Arias, 2009).

Although some *Bacillus* species are known to colonize human skin, *B. amyloliquefaciens* is normally only associated with plants, indicating that it was probably a transient bacteria on the cheek rather than a species of the normal microflora there. This is consistent with the findings from the initial swab of the cheek, which only yielded one colony of the bacteria that eventually was identified as *B. amyloliquefaciens*, as opposed to the four other colony types on the first plate, which were much more numerous.

Previous literature indicates that the *B. amyloliquefaciens* are Gram-positive rod-shaped bacteria that tend to form chains, as was indicated by the Gram stain and observations of cellular morphology of AJM-3 (Priest et al., 1987). The literature on colony morphology and coloring was fairly consistent with the present findings, with the exception that it described a circular form for the colonies rather than an irregular form (Yin et al., 2011). This may have been due to the subjectivity of the observation rather than an actual morphological discrepancy.

Our data (Table 1 [Table 8.4]) correlate well with results for testing methods described in the literature on *B. amyloliqefaciens*, which states that this species can hydrolyze starch, ferment lactose, and reduce nitrate (Welker and Campbell 1967). There is also correlation in the data stating that there is positive catalase and oxidase activity (Yin et al., 2011). However, the current study revealed that AJM-3 was negative for mannitol fermentation, which is not consistent with previous findings about *B. amyloliquefaciens* stating that the species does ferment mannitol as well (Priest et al., 1987).

Previous studies of antibiotic resistance in *B. amyloliquefacins* often focused on different antibiotics than those in this study. However, it was emphasized that this species is highly resistant to penicillin, a result that, as seen in Table 2 [Table 8.5], was reflected in the Kirby–Bauer test (Dias et al., 1986). There is also evidence of limited resistance to Streptomycin and Tetracycline, which was also observed in AJM-3 (Dias et al., 1986).

The growth curve data in a study by Maubert et al. is somewhat consistent with the growth curve of AJM-3 (2007). It appeared to enter into log phase around 240 min after incubation started and continued until around 480 min, which is a shorter log phase than AJM-3 exhibited. However, the conditions of growth were different, and overnight culture was not added to the broth as it was to the culture of AJM-3, making it difficult to objectively compare the two sets of data (Maubert et al., 2007). The strain may also be different,

Discussion compares and contrasts this study with others in the field.

which could potentially alter the growth curve. No generation time data were available for comparison for *B. amyloliquefaciens*, making the growth curve data least diagnostic in determining the correct species identification.

Finally, the Biolog data (Figure 2a [Figure 8.4a]) were compared to those of previous studies, revealing many consistencies, including positive carbon source utilization of Cellobiose, Maltose, Fructose, Mannose, Lactose, Maltose, Glycerol, Sorbitol, Raffinose, Salicin, Trehalose, Gelatin, and Tween 40 (Priest et al., 1987). Another notable consistency was the ability of AJM-3 (Figure 2b [Figure 8.4b]) and *B. amyloliquefaciens* to tolerate an environment of NaCl concentrations ranging from 1% to 8% (Priest et al., 1987).

Inconsistencies and limitations of the study are mentioned and discussed.

One identifiable inconsistency was with the use of pectin as a carbon source. The Microlog plate indicated that it was used as a carbon source by AJM-3, identified as *B. amyloliquefaciens*, while the literature indicates that pectin may not be utilized by this species (Priest et al., 1987). This may be because different strains of the same species of bacteria may have slight variations in metabolic capabilities, or, potentially, just because of contamination of that well.

Another inconsistency found was between the MicroPlate data and the data from the Mannitol Salt Agar Test; the MicroPlate data showed that the bacteria can use mannitol as a carbon source in fermentation, which is contradictory to the negative finding in the earlier testing. This discrepancy is most likely attributed to misreading the MSA plate, which could have occurred if the media color associated with an adjacent bacteria on the plate was mistaken for the media color surrounding AJM-3, given that the colors blended together somewhat on the plate.

Although there are a few small inconsistencies in morphology and metabolic capabilities, the high correlation between the testing results for AJM-3 and the established literature on *B. amyloliquefaciens* indicates that they are the same species. The inconsistencies are more than likely due to testing error or subjectivity rather than actual differences between the two species.

Last paragraph summarizes main findings and provides the conclusion and significance.

Since the species identification appears to be correct, it is interesting to note that the bacteria were located on the skin of the face, an area very different from where it normally colonizes. This indicates that the bacteria are transient and that the skin may have somehow come in contact with a plant or the soil. Since the hands are a part of the skin constantly exposed to different environments and surfaces, it is possible that the soil and plant bacterium was first transmitted to the hands and then transmitted to the skin on the cheeks when the hand came in contact with the facial skin at some point. These findings also indicate how bacteria can be transmitted to an area of the body unintentionally, reinforcing the need for people to be aware of how harmless or harmful bacteria can colonize an area of the body if we are not mindful of what contaminated environments or surfaces we come in contact with unknowingly. *Bacillus amyloliquefaciens* is not a human pathogen, but its presence on the human skin shows how bacteria can be easily transmitted from one environment to another on the human body under the right conditions.

Literature Cited

References are listed alphabetically.

Arguelles-Arias A, Ongena M, Halimi B, Lara Y, Brans A, Joris B, Fickers P. 2009. *Bacillus amyloliquefaciens* GA1 as a source of potent antibiotics and other secondary metabolites for biocontrol of plant pathogens. *Microb Cell Fact.* 8:63.

Chiller K, Selkin BA, Murakwaa GJ. 2001. Skin microflora and bacterial infections of the skin. *The Journal of Investigative Dermatology* 6(3):170–174.

Dias FFM, Shaikh KG, Bhatt YB, Modi DC, Subramanyam VR. 1986. Tunicamycin-resistant mutants *Bacillus amyloliquefaciens* are deficient in amylase, protease and penicillinase synthesis and have altered sensitivity to antibiotics and autolysis. *Journal of Applied Bacteriology* 60:271–275.

Authors are listed by last name, followed by first name initial(s) followed by year of publication, title of article, name of journal, volume and page number.

Edlund C, Nord CE. 2000. Effect on the human normal microflora of oral antibiotics for treatment of urinary tract infections. *Journal of Antimicrobial Chemotherapy* 46:41–48.

Evans CA, Smith WM, Johnson EA, Giblett ER. 1950. Bacterial flora of the normal human skin. *The Journal of Investigative Dermatology* 15:305–324.

Idriss EE, Makarewicz O, Farouk A, Rosner K, Greiner R, Bochow H, Richter T, Borriss R. 2002. Extracellular phytase activity of *Bacillus amyloliquefaciens* FZB45 contributes to its plant-growth-promoting effect. *Microbiology* 148:2097–2109.

Madigan MT, Martinko JM, Dunlap PV, Clark DP. 2009. *Brock Biology of Microorganisms*, 12th ed. Reading: Benjamin Cummings.

Marples RR. 1974. The microflora of the face and acne lesions. *The Journal of Investigative Dermatology* 62(3):326–331.

Maubert ME, Hartz CB, Willson K. 2007. Identification of a *Bacillus amyloliquefaciens* (BA) strainable to bioremediate methyl tertiary butyl ether (MTBE) *in vitro* and *in situ*. *Journal of Young Investigators* 17(1).

Priest FG, Goodfellow M, Shute A, Berkeley RCW. 1987. *Bacillus amyloliquefaciens* sp. nov., nom. rev. *International Journal of Systematic Bacteriology* 37(1):69–71.

Sullivan A, Edlund, C, Nord CE. 2001. Effect of antimicrobial agents on the ecological balance of human microflora. Review. *The Lancet: Infectious Diseases* 1:101–114.

Tancrede C. 1992. Role of human microflora in health and disease. Review. *European Journal of Clinical Microbiology and Infectious Disease* 11(11):1012–1015.

Welker NE, Campbell LL. 1967. Unrelatedness of *Bacillus amyloliquefaciens* and *Bacillus subtilis*. *Journal of Bacteriology* 94(4):1124–1130.

Willey JM, Sherwood LM, Woolverton CJ. 2008. *Microbiology*. 8th ed. New York: McGraw-Hill.

Yin X, Xu L, Xu L, Fan S, Liu Z, Zhang X. 2011. Evaluation of the efficacy of endophytic *Bacillus amyloliquefaciens* against *Botryosphaeria dothidea* and other phytopathogenic microorganisms. *African Journal of Microbiology Research* 5(4):340–345.

Books are also included in the reference list.

CHAPTER 9

Reading, Summarizing, and Critiquing a Scientific Research Article

9.1   CONTENT OF A SCIENTIFIC RESEARCH ARTICLE

GUIDELINE:

Understand what information to look for and where to find it in a research article.

If you are majoring in a scientific field, you will be reading articles published in academic and professional journals at some point. You might read these articles as part of a literature review you are required to submit, or your instructor may even ask you to write a summary or a critique of an article. If you are conducting research in a laboratory, you will definitely need to understand scientific research articles, which are also referred to as "research papers" (see also Chapter 6, Section 6.1). Therefore, it is essential that you learn how to understand such articles and how to summarize their content in your own words.

Research articles may seem overwhelming, especially to those who have little or no experience reading or writing this type of paper. Most likely you will not understand every detail. Reading and understanding "primary articles" or research papers is mainly a matter of experience and knowledge of the specific vocabulary of a field. However, a few simple tactics make this process much easier. Most important in this respect is to know what essential information you need to find and where to find this information in a research article.

Essential information you need to glean from an article includes:

- The overall purpose of the research
- The general experimental approach

- The major results
- The significance of the work

To know where to find this information, you need to understand how journal articles are structured (see also Chapter 6):

Abstract
- Provides a mini-summary of the paper and should contain all important information. The abstract will let you know what information the article covers.

Introduction
- Gives relevant background information and introduces the problem/unknown in the field.
- Last paragraph—states the overall question or purpose of the paper and the general approach. (In addition, it may also state the main results and their significance.)

Materials and Methods
- Tells the experimental approach. Most of the time this section is very technical and detailed. For an overview of the experimental approach, you can also refer to the end of the Introduction or to the Results section.

Results
- Delineates all findings of the work .
- First paragraph(s)—states the main results of the research.
- Subsequent paragraphs—outline the general approach and findings for each step of the study.

Discussion
- Discusses the key findings and their significance.
- First paragraph(s)—informs about the main results and their meaning .
- Last paragraph/conclusion—points out the importance and potential impact of the research; may also restate the main findings and point to the direction of future studies.

9.2   READING A RESEARCH PAPER

GUIDELINE:
Read research papers in a directed way.
Gain an overview first.
Clarify questions and unfamiliar terminology.
Take notes.

Given knowledge of the general layout of a paper, you can read research articles in a directed way:

1. Read the abstract, the introduction, and the conclusion. These sections should provide you with the overall purpose of the paper, the reason why the problem is important, the main findings, and the overall importance of the paper.
2. Read through the entire paper. Look especially closely at the figures. Write down questions you have and terminology you do not understand. Look up simple words and phrases. This may help you answer many, if not most, of your questions. A medical or biological dictionary or the internet is a great source for definitions. A textbook of the specific field may also be a good source, because it gives more complete explanations.
3. Re-read the paper again—this time, for fuller comprehension. If you need to write a summary or critique, take notes on each section using the questions in Section 9.3 or 9.4 as guidelines.

This approach should provide you with a solid idea of what this paper is about, what the main results are, and why they are important. You will be able to summarize or critique material only if you understand these components, and you can write a successful term paper, literature review, or even your own publication only if you can summarize your research material.

9.3   WRITING A SUMMARY OF A RESEARCH PAPER

GUIDELINE:

Think of a summary as an expanded version of the abstract, written in your own words.

Summarizing (or critiquing) a professional paper is a great way to learn more about primary scientific articles, writing research papers, and the research process itself. You will learn how researchers conduct experiments, interpret results, and discuss their impact. Writing these summaries will also help you distinguish between more and less important material within a paper.

After reading the article carefully, specific questions you should be able to answer include:

- What is the overall purpose of the research?
- How does the research fit into the context of its field?
- What was the general experimental approach?
- What are the key findings?
- How are the reported findings different or better?
- What are the major conclusions drawn from the findings?
- What is the overall importance of the work?

To write a good and concise summary, think of your summary as an expanded version of the abstract of the article in your own words. The following outline can serve as a guideline:

1. Begin your summary by describing the main question or purpose of the paper and provide some brief context.
2. Explain how the authors approached the study.
3. Describe the key findings.
4. Briefly discuss the meaning of the findings.
5. Conclude by stating the overall impact of the research and explain why you think the study is relevant.

Do not feel obligated to summarize each detail of the paper. Focus instead on giving the reader an overall idea of the content of the article. Ensure that the terminology you are using is correct, and limit your summary to roughly one page.

Example 9-1

Summary of "Effects of Mountains and Ice Sheets on Global Ocean Circulation" *

The article "Effects of Mountains and Ice Sheets on Global Ocean Circulation" describes the impact of mountains and ice sheets on the large-scale circulation of the world's oceans. The researchers used a series of simulations with a new coupled ocean–atmosphere model [Oregon State University–University of Victoria model (OSUVic)], which combines ocean, sea ice, land surface, and ocean biogeochemical model components, to analyze the orographic effects of mountains and ice sheets on zonal wind stress, surface fresh water budgets, precipitation, atmospheric circulation, air temperature, and eddy kinetic energy.

The investigations show that the higher the mountain ranges, the less water vapor is transported from the Pacific into the Atlantic Ocean. As a result, the Atlantic is saltier and the Pacific is fresher because of the Rocky Mountains and Andes. In addition, deep-water formation is increased in the Atlantic but decreased in the Pacific, leading to more stratification in the Pacific.

Orography also affects zonal wind stress. Higher mountain ranges lead to increased zonal wind stress in the Southern Ocean, which is shifted southward by the presence of the Antarctic ice sheet. The combination of less atmospheric water vapor export from the Pacific to the Atlantic and southward-shifted Southern Hemisphere westerlies enhance deep-water formation in the North Atlantic and accelerate the Atlantic meridional overturning circulation.

The authors conclude that the configuration of mountains and ice sheets is key to determining the observed global circulation pattern of the ocean. Their findings may be important for future changes in this pattern due to climate changes—and consequently, important consequences for the carbon cycle—particularly if the Antarctic ice sheet is reduced or disappears.

*(Smittner et al., 2011, J. Climate, **24**, 2814–2829)

9.4   CRITIQUING A RESEARCH PAPER

GUIDELINE:

Highlight the strengths and weaknesses of the research, its presentation, and interpretation.

In a critique you are evaluating work done by another author. The purpose is to highlight strengths and weaknesses of the research approach, presentation, or interpretation in one to two pages. At the same time, you will deepen your understanding of the paper.

In contrast to a summary, composing a research paper critique is subjective writing. It allows you to point out shortcomings and to suggest ways of improvement for the research, its presentation, or its interpretation. For example, in the summary you would answer the question *"What are the major conclusions drawn from the findings?"*, whereas in a critique you would need to look at the conclusions critically, asking instead *"Are the major conclusions drawn from the findings justified?"*

Similar significant questions that you need to answer when writing a research paper critique are listed in the following table, contrasted to those of a simple summary:

SECTION	SUMMARY	CRITIQUE
Introduction	• What is the overall purpose of the research? • How does the research fit into the context of its field?	• Is the purpose of the study clearly stated? • Are the ideas novel/original? • Has relevant background information been provided?
Methods	• What was the general experimental approach?	• Did the authors use appropriate measurements and procedures? Do they help to answer the question of the paper?
Results	• What are the key findings?	• Did the authors observe the expected results? • Are results correctly interpreted and were all controls met? • If some assumptions are made, are they realistic? • Are figures and tables explained clearly?
Discussion	• How are the reported findings different or better?	• Do the authors answer the overall research question? • Did the authors make clear how the research fits into the context of its field? • Does the work make an important contribution to the field?
Conclusions	• What are the major conclusions drawn from the findings?	• Are the major conclusions drawn from the findings justified? • What would you add to the conclusion? • What would you say differently?
Importance	• What is the overall importance of the work?	• Are the contributions significant?

These guiding questions can also be helpful in learning to critique your own research paper.

Example 9-2

Critique of "Effects of Mountains and Ice Sheets on Global Ocean Circulation" *

In the article "Effects of Mountains and Ice Sheets on Global Ocean Circulation," two features influencing the large-scale circulation of the world's oceans are investigated: mountains and ice sheets. Large-scale ocean circulation, which contributes to our climate system, is influenced by wind, buoyancy forcing, momentum fluxes, and orographic effects, such as mountain configurations and ice sheets. The latter two features have not been investigated in any detail prior to this study.

To assess the impact of orography, the researchers used a novel coupled ocean–atmosphere model [Oregon State University–University of Victoria model (OSUVic)], which combines ocean, sea ice, land surface, and ocean biogeochemical model components. Their work analyzes the orographic effects of mountains and ice sheets on zonal wind stress, surface fresh water budgets, precipitation, atmospheric circulation, air temperature, and eddy kinetic energy through four experimental calculations.

Their modeling shows that the higher the mountain ranges, the less water vapor is transported from the Pacific into the Atlantic Ocean. As a result, the Atlantic is saltier and the Pacific is fresher and deep-water formation is increased in the Atlantic but decreased in the Pacific. In addition, higher mountain ranges lead to increased zonal wind stress in the Southern Ocean, which is shifted southward by the presence of the Antarctic ice sheet. The authors conclude that the configuration of mountains and ice sheets is key to determining the observed global circulation pattern of the ocean.

The study does not take into account the influence of fresh water streams and currents on the salinity of the oceans nor gravity wave grads or any anthropogenic changes such as depletion of the ozone in certain areas. Furthermore, as they point out themselves, their horizontal resolution is relatively coarse, consisting of only 10 layers. Further simulations considering these additional factors and using a state-of-the-art climate model and a finer resolution would therefore be desirable. Without question, their findings may be important for future changes in the ocean circulation pattern due to climate changes, particularly if the Antarctic ice sheet is reduced or disappears.

(Smittner et al., 2011, J. Climate, **24, 2814–2829)*

SUMMARY

WORKING WITH A RESEARCH PAPER

1. Understand what information to look for and where to find it in a research article.

2. Read research papers in a directed way.
   - Gain an overview first.
   - Clarify questions and unfamiliar terminology.
   - Take notes.
3. Think of a summary as an expanded version of the abstract written in your own words.
4. To critique a research paper, highlight the strengths and weaknesses of the research, its presentation, and its interpretation.

CHAPTER 10

Term Papers and Review Articles

10.1   GENERAL ADVICE

Review articles knit together theories and results from a number of studies to provide an overview or evaluation of a field of research. Thus, they are not original articles with new data but, instead, secondary sources representing a well-balanced summary of a timely subject. (See also Chapter 2, Section 2.1 and 2.2, for discussion on primary, secondary, and tertiary sources.)

Term papers or reviews, which upper-level undergraduate and graduate students are often asked to write, can be viewed as a simplified version of a review article, the difference being that review papers deal more in depth with current topics published in the literature and are longer than term papers. Review articles similarly outline the overall picture of a particular topic as it is currently understood by scientists in that field, but go into much more depth than term papers.

The emphasis in both types of papers is on interpretation and evaluation. Unlike a "book report" style summary with which you may be familiar from grade school, a good term paper or review article is one that furthers our understanding of or adds a new dimension to a topic. By comparing and contrasting key studies done in a certain research area, term papers and review articles analyze how each line of research supports or fails to support a theory. Such evaluations provide scientists with the most up-to-date information as well as with the history and a critical evaluation of the topic. These papers also provide a comprehensive knowledge of the literature in the field. In addition, term papers and review articles outline any problems that are currently being addressed and explain the basis of any conflicts that exist between experts in the field. As the author of such an article, you can suggest which side of the conflict seems to be presenting the better arguments. You can also suggest possible next steps or propose a new model.

It is particularly important that the information in a term paper or review article is understandable to scientists in other related fields. Thus, reviews should use simple words and avoid excessive jargon and technical detail. Figures and boxed material should summarize and generalize primary source data or highlight new ideas.

10.2   SOURCE MATERIAL

Some research topics are much easier to write about than others. In writing a review, it is wise to choose a topic of current interest and to pick a research topic about which articles are continuing to be published. Subjects of well-defined and well-studied areas of research typically give more fruitful topics.

To find out what is "hot" or "cutting edge," it is often helpful to read a couple of review articles from a variety of journals. Other literature reviews in your area of interest may also give you a sense of the types of themes you might want to look for in your own research or may suggest ways to organize your final review. Other ways to identify active areas of research are through reading editorials and letters to the editor.

There are two main approaches to choosing an area of research to write about in a term paper or review article. One approach is to choose a point that you want to make and then select your primary studies based on this area of interest. Another approach is to read all the relevant studies and organize them in a meaningful way. That is, research the topic starting at general sources and work your way to specific sources. Begin by looking at textbooks or internet sources that are vetted through peer review or have refereed references backing up statements. These are generally considered to be tertiary sources and can provide a good overview of a topic. Next, narrow your search by reading up on topics that have been summarized in secondary sources such as in review articles. For both approaches, use secondary sources to come up with a general skeleton for your review paper. Then, use gathered, specific information from primary sources, which are first-hand accounts of investigations and include journal articles, theses, and reports, to fill in the skeleton of your outline.

For all your sources, keep good records from the beginning, particularly citation information. There is nothing more aggravating than having to rediscover where the source material originated from.

10.3   ORGANIZATION

GUIDELINE:

Follow the overall structure
Title
Abstract—not always required
Introduction
Main analysis section

Conclusion and/or recommendations
Acknowledgments (only for articles for publications)
References

The standard organization of term papers or review articles differs from that of lab reports or research papers. The organization of term papers and review articles usually includes an introduction to the topic, a main section with headings and subheadings, a conclusion with recommendations for further research, and a lengthy reference section. Some term papers and review articles also contain an abstract. Depending on what type of article you write, your paper could follow a slightly different organization.

To compose a logically structured review article, create an outline. If you decide to compose your article without one, at least check the overall organization using a reverse outline during the revision stage. For a clear outline of your term paper or review paper, create subsections based on the information you have gathered from the literature. Give these subsections individual headings and subheadings, and then sort the information you have collected into the various subsections. Use bullet points or whole sentences under each heading or subheading. Subsequently, sort the information under each heading or subheading by similarities, contrasts, gaps in knowledge, and so forth.

As you are filling in your outline, reread the source articles to ensure that you have not missed anything. Identify additional papers if needed, and re-sort your material again if necessary. Writing a review article is an iterative process. When you are satisfied with your outline, start writing the review article by linking all the ideas under each subheading. The following sections provide more details on how to logically organize the Introduction, Main Analysis section, and Conclusion of a review article.

In Example 10-1, an outline of a review is presented:

 Example 10-1   Outline of Review Article

Title: Species-specific cell recognition: Gamete adhesion of sea urchins
Abstract—see Section 10.4
Introduction
　　Cell adhesion—overview
　　Cell adhesion model system—sea urchins
　　Advantages of model system
　　Gamete interactions in two model species
Main Analysis
　　Macromolecular interactions in sea urchin fertilization
　　Species specificity in fertilization
　　Bindin and sperm adhesion
　　　Support for bindin's function
　　　　Evidence for molecular interactions
　　　　Reports on bindin's receptor

Deducted hypothesis: model of interaction between bindin
and its receptor
  Evidence through deletion mutants
  Evidence through sequence analysis
Conclusion
  Summary of main findings
  Summary of hypothesis
  Projection to other species
Acknowledgments
References

In this outline, the authors progress from a general overview of cell adhesion to specific gamete interactions in sea urchin fertilization in the Introduction. For the Main Analysis section, the article is logically organized into discussions of macromolecular interactions in sea urchin fertilization, the species specificity of these interactions, and the molecular interactions of sperm adhesion. The Main Analysis section ends with a proposed model for the interaction of the specific sperm and egg adhesive molecules based on the collective evidence from previous studies presented in prior subsections. This hypothesis fills the gap in knowledge of the exact molecular interactions during sea urchin fertilization. The outline concludes by summarizing the main findings and resulting hypothesis and by projecting the possibility of similar interactions during fertilization in other species.

10.4   ABSTRACT

GUIDELINE:

Write the abstract as a table of contents in paragraph form.

Structure the abstract as follows:

- Background (optional)
- Problem statement (optional)
- **Purpose/topic of review**
- **Overview of content**

Not all review articles require an abstract. If they do, their abstracts differ from those of a research paper. Unlike abstracts for research papers, those for term papers and review articles are essentially tables of contents in paragraph form. They may contain some background information and/or a problem statement and should state the purpose or topic of the review. They usually include little, if any, methods or results. Although some abstracts end with a statement of significance, interpreting the main findings of a topic for the readers, most end with an overview sentence, listing what will be discussed in the document. Note, however, that such overview sentences are not appropriate in abstracts of lab reports or scientific research articles.

Example 10-2   Abstract of review article

Overview of content	This paper describes how plants use the structural diversity of oligosacchrides to regulate important cellular processes such as growth, development, and defense. **We address the central remaining question of how cells perceive and transduce oligosaccharide signals and discuss current research aimed at providing the answer.**

Longer abstracts for review articles contain the following elements: background, problem statement, statement of topic, and overview of content.

Example 10-3   Abstract of review article

Background	Aerosols serve as cloud condensation nuclei (CCN) and thus have a substantial effect on cloud properties and the initiation
Unknown/ Problem	of precipitation. Large concentrations of human-made aerosols have been reported to both decrease and increase rainfall as a result of their radiative and CCN activities. At one
Purpose/topic statement	extreme, pristine tropical clouds with low CCN concentrations rain out too quickly to mature into long-lived clouds. On the other hand, heavily polluted clouds evaporate much of their water before precipitation can occur, if they can form at all given the reduced surface heating resulting from the aerosol
Overview of content	haze layer. We propose a conceptual model that explains this apparent dichotomy.

(With permission from the American Association for the Advancement of Science)

10.5   INTRODUCTION

GUIDELINE:

Organize the Introduction

- Background
- Unknown or problem
- Purpose/topic of review
- Overview of content

The introduction of a term paper or review article should provide the big picture of the topic and grab the readers' attention. It should present some general background and state the central purpose/topic of the review. It should also make clear why the topic warrants a review.

After discussing the general background and aspects of existing research, the article should present any recent developments and describe what the problems with the existing research are and/or what is unknown. Subsequently, it should explain the overall purpose of the review article.

This statement should be followed by a description of its organizational pattern. Do not make the introduction longer than one-fifth of the review article.

The statement of purpose/topic of a term paper or review article is similar to the question or purpose of a research paper. Your statement of purpose/topic will not necessarily argue for a position or an opinion; rather, it will argue for a particular perspective on the topic. The statement tells the reader how you will interpret the significance of the subject matter under discussion and lets the reader know what to expect from the rest of the article. Sample statements of purpose/topic for literature reviews include

---

Example 10-4   Purpose/Topic statement for review papers

The development of new antibiotics has become the main focus of several biotech companies.

Mathematical modeling of disease transmission is important to maximize the utility of limited resources.

In the following example of a complete introduction of a review article the various components are indicated.

Example 10-5   Introduction of a review paper

Background | Global climate change is altering the geographic ranges, behaviors, and phenologies of terrestrial, freshwater, and marine species. A warming climate, therefore, appears destined to change the composition and function of marine communities in ways that are complex and not entirely predictable (1-5). Higher temperatures are expected to increase the introduction and establishment of exotic species, thereby changing trophic relationships and homogenizing biotas (6). Because

Unknown/ Problem | organisms in polar regions are adapted to the coldest temperatures and most intense seasonality of resource supply on Earth (7), polar species and the communities they comprise are especially at risk from global warming and the concomitant invasion of species from lower latitudes (8–10).

Background | Shallow-water, benthic communities in Antarctica (<100-m depth) are unique. Nowhere else do giant pycnogonids, nemerteans, and isopods occur in shallow marine environments, cohabiting with fish that have antifreeze glycoproteins in their blood. An emphasis on brooding and lecithotrophic reproductive strategies (11, 12) and a trend toward gigantism (13) are among the unusual features of the invertebrate fauna. Ecological and evolutionary responses to cold temperature underlie these peculiarities, making the Antarctic bottom fauna particularly vulnerable to climate change. The Antarctic benthos, living at the lower thermal limit to marine life, serves as a natural laboratory for understanding the impacts of climate change on marine systems in general.

| Purpose/
Topic statement | Recent advances in the physiology, ecology, and evolutionary paleobiology of marine life in Antarctica make it possible to predict the nature of biological invasions facilitated by global warming and the likely responses of benthic communities to such invasions. This review draws on paleontology, biogeography, oceanography, physiology, molecular ecology, and community ecology. We explore the climatically driven origin of the peculiar community structure of modern benthic communities in Antarctica and the macroecological consequences |
| Overview | of present and future global warming. |

(With permission from Annual Reviews)

10.6   MAIN ANALYSIS SECTION

GUIDELINES:

Organize the Main Analysis section logically into subsections either

- chronologically
- thematically
- methodologically

Logically organize information within the Main Analysis subsections (similarities, contrasts, gaps in knowledge, etc.)

One of the most difficult tasks in writing a term paper or review article is finding the best structure for the Main Analysis section. This section should present any experimental evidence by describing important results from recent primary literature articles and explain how these results shape current understanding of the topic while addressing any controversies in the field. You may use figures and/or tables to present your interpretation of original data or show key data taken directly from the original papers. Ensure that you are not plagiarizing, however. Cite sources of information as needed. (See also Chapter 2, Section 2.6.)

Generally, this section of a term paper or review article can be organized chronologically, thematically, or methodologically. If your review follows the chronological method, you could write about topics according to when they were published. Alternatively, you could examine the sources under the history of the topic. Such an organization would call for subsections according to eras within this history.

In contrast to the chronological presentation of topics, thematic reviews are organized around a topic or issue rather than around the progression of time. For example, as you deal with various levels of evidence pertaining to a question, your Main Analysis could move steadily downward in the level of inquiry from the organism, to the organ, to the cell, to the molecular mechanisms within the cell. However, progression of time may still be an important factor in a thematic review.

A methodological approach differs from the preceding two in that its focus usually does not have to do with the content of the material. Instead, it focuses on the "methods" of the researchers, and topics are organized accordingly by techniques or by methods or approaches.

Once you have decided on the organizational method for the main section of the review, the subsections you need for the review paper should arise out of your organizational strategy. In other words, a chronological review would have subsections for each vital time period, whereas a thematic review would have subtopics based on factors that relate to the theme or issue.

For some reviews, you might need to add additional sections that are necessary for your study but do not fit in the organizational strategy of the main section (e.g., "Current Status" or "Future Directions"). What subsections you include in the body may only become clear as the review evolves.

In the next example, a partial Main Analysis subsection of a review article is shown.

Example 10-6   Main analysis section of a review paper

Subheading ____ **The Opposing Effects of Aerosols on Clouds and Precipitation**

With the advent of satellite measurements, it became possible to observe the larger picture of aerosol effects on clouds and precipitation. (We exclude the impacts of ice nuclei aerosols, which are much less understood than the effects of CCN aerosols.) Urban and industrial air pollution plumes were observed to completely suppress precipitation from 2.5-km-deep clouds over Australia (20). Heavy smoke from forest fires was observed to suppress rainfall from 5-km-deep tropical clouds (21, 22). The clouds appeared to regain their precipitation capability when ingesting giant (>1-μm diameter) CCN salt particles from sea spray (23) and salt playas (24). These observations were the impetus for the World Meteorological Organization and the International Union of Geodesy and Geophysics to mandate an assessment of aerosol impact on precipitation (19). This report concluded that "it is difficult to establish clear causal relationships between aerosols and precipitation and to determine the sign of the precipitation change in a climatological sense. Based on many observations and model simulations the effects of aerosols on clouds are more clearly understood (particularly in ice-free clouds); the effects on precipitation are less clear."

A recent National Research Council report that reviewed "radiative forcing of climate change" (25) concluded that the concept of radiative forcing "needs to be extended to account for (1) the vertical structure of radiative forcing, (2) regional variability in radiative forcing, and (3) nonradiative forcing." It

Context

Problem/
Unknown

Proposed
solution

<table>
<tr><td>Proposed solution</td><td>recommended "to move beyond simple climate models based entirely on global mean top of the atmosphere radiative forcing and incorporate new global and regional radiative and nonradiative forcing metrics as they become available." We propose such a new metric below.</td></tr>
</table>

(With permission from the American Association for the Advancement of Science)

10.7 CONCLUSION

GUIDELINE:

In the conclusion section, summarize your topic, generalize any interpretations, and provide some significance.

The conclusion section is one of the main highlights of a term paper or review paper. It recaps your review and your main conclusions, recommendations, and/or speculations.

You need to phrase this section with special care, summarizing and generalizing main lines of arguments and key findings. Discuss what conclusions you have drawn from reviewing the literature, and restate your interpretations. In addition, provide some general significance of the topic and results, and discuss the questions that remain in the area. Although this section is often longer than the conclusion section of a research paper, try to keep it brief.

Example 10-7 provides an example of a well-written Conclusion section for a review article.

Example 10-7   Conclusion of a review paper

Beginning in the late Eocene, global cooling reduced durophagous predation in Antarctica. Despite some climatic reversals, the post-Eocene cooling trend drove shallow-water, benthic communities to the retrograde, Paleozoic-type structure and function we see today. Now, global warming is facilitating the return of durophagous predators, which are poised to eliminate that anachronistic character and remodernize the Antarctic benthos in shallow-water habitats. Rising sea temperatures should in general act to reduce the mismatch between the development times of invasive larvae and the length of the growing season. Increased survivability of planktotrophic larvae will decrease the selective advantage of brooding and lecithotrophy, increasing the pool of potentially invasive species. Warming temperatures will also increase the scope for more rapid metabolism and should ultimately obviate the adaptive value of gigantism. All of these effects appear destined to amplify the ongoing, worldwide homogenization of marine biotas by reducing the endemic character of the Antarctic fauna.

Summary

| | The fact that benthic predators are already beginning to invade the Antarctic Peninsula should be taken as an urgent warning. Controlling the discharge of ballast water from ships will be difficult but not impossible. Whether or not humans are the proximal vectors, however, the long-term threat of invasion in Antarctica has its roots in climate change. The Antarctic Treaty cannot control global warming. Global environmental policy must immediately be directed to reducing and reversing anthropogenic emissions of greenhouse gases into the atmosphere if marine life in Antarctica is to survive in something resembling its present form. |
| Interpretations and recommendations | |

(With permission from Annual Reviews)

10.8   REFERENCES

Like in academic research papers, your interpretation of the available sources must be backed up with evidence. Therefore, cite primary and secondary sources where needed. The type of information you choose should relate directly to the review's focus. See Chapter 2 for more information on references.

10.9   CHECKLIST

Overall

☐   Does the topic present something of interest to the field?

Individual sections

Abstract:

☐   Is the purpose or topic stated precisely?

☐   Is an overview of the article provided?

Introduction:

☐   Does the Introduction have the following components?

☐   Background

☐   Problem or unknown

☐   Purpose/topic or review

☐   Overview of content

Main Analysis Section:

☐   Is your section logically organized?

☐   Did you analyze and interpret all information (rather than simply list facts and dates)?

☐   Does your paper present information objectively, including contradictory data and ambiguities?

☐   Are all figures and tables explained and labeled sufficiently?

☐   Do all the components logically follow each other?

Conclusion and/or Recommendations:

☐   Is the topic summarized and interpreted in the Conclusion section?

☐   Is the significance of your analysis clear?

References:
- ☐ Have references been cited where needed?
- ☐ Are sources cited adequately and appropriately?
- ☐ Are all the citations in the text listed in the References section?

STYLE AND COMPOSITION
- ☐ Are the transitions between sections and paragraphs logical?
- ☐ Are paragraphs consistent?
- ☐ Are paragraphs and sentences cohesive?
- ☐ Has word location been considered?
- ☐ Has grammar, punctuation, and spelling been checked?
- ☐ Is the style concise?
- ☐ Are key terms consistent?
- ☐ Is the action in the verbs? Are nominalizations avoided?
- ☐ Did you vary sentence length and use one idea per sentence?
- ☐ Are comparisons written correctly?
- ☐ Are lists parallel?
- ☐ Have noun clusters been resolved?
- ☐ Are nontechnical words and phrases simple?
- ☐ Have unnecessary terms (redundancies, jargon) been reduced?

SUMMARY

A term paper or review article should follow the following overall structure:

1. Abstract:
   - Content: Background (optional), Problem statement (optional), Purpose/topic of review, Overview of content

2. Introduction: Organize into funnel shape
   - First paragraphs: Background
   - Second to last paragraph: Unknown or problem
   - Last paragraph: Purpose/topic of review and Overview of content

3. Main Analysis Section:
   - Organize logically into subsections either chronologically, thematically, or methodologically
   - Logically organize information within the subsections (similarities, contrasts, gaps in knowledge, etc.)

4. Conclusion Section:
   - Summarize the topic, generalize any interpretations, and provide some significance.

PROBLEMS

Problem 10-1 Review Introduction

Identify all the elements of the following review introduction? Is the introduction complete?

Mitochondrial genomes differ greatly in size, structural organization and expression both within and between the kingdoms of eukaryotic organisms. The mitochondrial genomes of higher plants are much larger (200–2400 kb) and more complex than those of animals (14–42 kb), fungi (18–176 kb), and plastids (120–200 kb) (Refs. 1–4). Although there has been less molecular analysis of the plant mitochondrial genome structure in comparison with the equivalent animal or fungal genomes, the use of a variety of approaches—such as pulsed-field gel electrophoresis (PFGE), moving pictures (movies) during electrophoresis, restriction digestion by rare-cutting enzymes, two-dimensional gel electrophoresis (2DE) and electron microscopy (EM)—has led to substantial recent progress. Here, the implication of these new studies on the understanding of *in vivo* organization and replication of plant mitochondrial genomes is assessed.

(With permission from Elsevier)

Problem 10-2 Review Abstract

Is the following abstract an abstract of a research article or that of a review article? Explain why.

Interleukin 1 (IL-1), a cytokine produced by macrophages and various other cell types, plays a major role in the immune response and in inflammatory reactions. IL-1 has been shown to be cytotoxic for tumor cells. To determine the effect of macrophage-derived factors on epithelial tumor cells, we cultured human colon carcinoma cells (T84) and an intestinal epithelial cell line (IEC 18) with purified human IL-1. Microscopic and photometric analysis indicated that IL-1 has a cytotoxic effect on colon cancer cells as well as cytotoxic and growth inhibitory effects on intestinal epithelial cells. Because IL-1 is known to be released during inflammatory reactions, this factor may not only kill tumor cells but also affect normal intestinal cells and may play a role in inflammatory intestinal diseases.

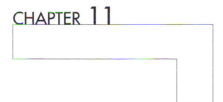

CHAPTER 11

Answering Essay Questions

11.1 GENERAL

Having to write an essay exam may inspire fear, particularly because this will involve time pressure to complete the task. However, essay test questions also provide an opportunity for you to present information that displays your understanding of the topic clearly. Accordingly, many instructors like essays because they challenge students to create a response rather than to simply select a response, showing their understanding of a subject matter as well as their abilities to reason. In addition, essays allow an instructor to assess writing abilities.

11.2 PREPARING FOR AN ESSAY EXAM

GUIDELINE:

Prepare for an essay exam.

The best and most important preparation for an essay exam is to study and understand the material. Memorize key events, facts, and names. Go to lectures, do the reading that is required, and learn the material as it is presented. Do not try to "cram" just before the exam. You will not be able to remember and know all the material at the depth required for an essay test by just trying to memorize as much as possible the night before.

You can prepare for an essay exam by:

- Reviewing lecture notes and supplementary material
- Organizing and prioritizing information into thematic patters
- Practicing answering questions on a study sheet if provided

- Studying review questions
- Discussing the review questions with other students to learn about how they are best answered
- Practicing writing essays by doing all homework involving essay questions
- Anticipating essay question(s) and practicing writing their answers. Consider, for example, turning the major boldface headings in your textbooks into questions using typical key words such as describe, explain, compare, and contrast.

11.3 ANSWERING AN ESSAY QUESTION

To answer essay questions successfully, you need to:

1. Read and understand the question.
2. Jot down notes/brainstorm and make an outline.
3. Write the first sentence as an overview sentence.
4. Logically organize and present your ideas.
5. Proofread your answer.

GUIDELINE:

Read and understand the question.

Read each question carefully. If the question is not clear to you, ask your instructor for clarification.

Questions consist of three important elements:

- **Subject**, which tells you what the general topic is
- **Limiting Words**, which tell you on what part of the topic your answer should focus
- **Direction Words**, which tell you what to do.

To present a relevant answer, you need to analyze the question based on these three elements. This analysis will also allow you to structure the Introduction for the answer.

 Example 11-1

> *Essay question: Select one aspect of the effects of solar flares on Earth and discuss it in depth in relation to one level of reasoning.*

Subject: effect of solar flares on Earth
Limiting words A: one aspect only of the Subject
Direction word 1: select . . .
Limiting words B: (your chosen aspect) in relation to one level of reasoning.
Direction word 2: discuss in depth . . .
Note: you are being asked to choose *one aspect only* of the subject, and discuss it in relation to *one level only* of reasoning.

Table 11.1 Direction Words

Define	Explain the meaning; often you are also expected to provide an appropriate example or two.
List/Identify/Outline	Write a list of the main points with brief explanations.
Summarize	Briefly state the main ideas in an organized way.
Describe/Illustrate	Present the main points with clear examples.
Discuss/Explain	Present the main points, facts, and details of a topic/give reasons.
Compare and Contrast	Present similarities and differences.
Contrast/Differentiate/ Distinguish	Present only the differences.
Criticize/Evaluate	Discuss the strengths and weaknesses/pros and cons/ advantages and disadvantages.
Argue	Present your opinion and defend it with examples.
Interpret	Present your analysis of the topic using facts and reasoning.
Justify/Prove	Present evidence and reasons supporting the topic.
Diagram/Draw/Illustrate	Make a drawing and label it.
Trace	Discuss according to a pattern such as chronologically.

In your essay answer, do not write everything you know about the topic or subject. Instead, focus on the limiting words. If you are asked to include a particular material in your assignment or to present a specific approach or angle (for example: Restrict your answer to altitudes of more than 1000 m), make sure you incorporate them in your answer. If your instructor has set a word limit for the answer of the essay question, you should not go over that limit to avoid losing points.

Pay particular attention to the direction words used in the question. You may find it helpful to circle or underline them. Learn to distinguish between the various direction words and what specifically they are asking you to do (see Table 11.1).

GUIDELINE:

Jot down notes/brainstorm and make an outline

After you have read and analyzed the question, write a brief outline of the answer. This will be particularly important for long essay questions. Making a quick diagram can also help you focus your thoughts.

Start by jotting down notes, then put the information in logical order— for example, chronological, most to least important, or pro and con. At this point, do not worry too much about proper sentence and paragraph construction, but be sure to write down all pertinent information. The logical organization of your ideas will be essential for earning a good grade.

Good extended-response answers have three parts: a beginning/introduction, a middle/body, and an ending/conclusion. In the introduction, present your main idea(s). In the body, provide information, examples, and details to support your main idea(s). If the answer calls for much information, you may need more than one paragraph here. In the conclusion, sum up your main idea(s).

Continuing on our prior Example, your notes may, for instance, read as follows:

Example 11-2

> *Essay question: Select one aspect of the effects of solar flares on Earth and discuss it in depth in relation to one level of reasoning.*
>
> **Notes:** Solar flares affect climate conditions on Earth.
> Examples of effects:
> - Increase in electromagnetic waves
> - Increase in temperature
>
> Increase in temperature:
> - mechanism
> - length
> - Acc to X theory, increase in regional temperature changes but will not affect overall climate change

---

GUIDELINE:

Write the first sentence as an overview sentence.

When you are ready to compose your answer, state your main point in the first sentence. This is best done by writing an overview sentence that repeats many of the words found in the question and provides an overview of the details to come. For our prior example, your opening sentence may read as follows:

Example 11-3

> *Essay question: Select one aspect of the effects of solar flares on Earth and discuss it in depth in relation to one level of reasoning.*
>
> **First sentence:** "Solar flares affect climate conditions on Earth in various ways."

To provide you with another question and a similar approach in answering it, here is another example:

Example 11-4

> *Essay question: "How do sperm cells of sea urchins and star fish differ?"*
>
> **First sentence:** "The sperm of sea urchins and star fish differ in three main ways."

A different approach in writing the first sentence of the answer to an essay question is to pick a working title for your essay. Your essay question may not ask for one, but a title can keep you focused and ensure that you answer the right question and that you answer the entire question. The title could be the topic, a category term, or the point of view that you intend to use. After you have come up with a working title, turn it into your topic statement by writing it as a complete sentence.

Example 11-5

> *Question:* "How do sperm cells of sea urchins and star fish differ?"
> **Working title:** "Sperm cells of marine invertebrates"
> **First sentence:** "The sperm cells of marine invertebrate differ in morphology and physiology according to phylogeny, spawning habitat, and fertilization strategy."

GUIDELINE:
Logically organize and present your ideas.

To answer test questions successfully, you need to be aware of the expected length of the essay. For short essay answers, you will have less time and space. It is usually best to start with a general statement and then move on to describe specific applications or explanations. For short answer essay questions, a paragraph will almost always suffice. Make sure that you write it in complete sentences.

Example 11-6

> *Question:* "How do sperm cells of sea urchins and star fish differ?"
> **Working title:** "Sperm cells of marine invertebrates"

First sentence	"The sperm cells of marine invertebrate differ in morphology and physiology according to phylogeny, spawning habitat and fertilization strategy. For example, most sea urchin sperm cells are blunt ended at the acrosomal tip and contain only a small amount of F-actin prior to the acrosome reaction. Their chemoattractant is resact, a 14-residue peptide. In contrast to sea urchin sperm cells, the sperm cells of starfish are more slender, have a much longer acrosomal tip with large amounts of F-actin, and have preformed acrosomal filaments. Their chemoattractant is startrak, a 13-kDa heat-stable protein. In addition, whereas the sperm cells of sea urchins are rather species-specific, those of starfish
Body	are not. The patterns of variation in sperm morphology and
Concluding sentence (optional)	physiology may explain the mechanisms by which species recognition evolves."

Extended-response essay questions take care and thought, but they are not complicated. Long-answer essays usually ask about several basic elements of a topic, or they ask you to compare and contrast two ideas. Answers to such questions should be three to five paragraphs in length and follow the following general format:

Introduction (first sentence or paragraph)
Body (1–3 paragraphs)
Conclusion (last paragraph)

To compose the body of your essay, use the bullet points you brainstormed and organized in your outline to construct the individual paragraphs of your essay (see also Chapter 4). Begin each paragraph with a topic sentence; then support the topic sentence with reasons and/or examples. Fill in the details, facts, and ideas necessary to support the points you have developed in your outline. Create good flow by connecting sentences either through placement of information by considering word location (see Chapter 3 and 4), repetition of key terms (see Chapter 4), or use of transitions (such as *first, second, next, finally, on the other hand, consequently, furthermore, in conclusion;* see Chapter 4).

Ensure that the information within your paragraphs is logically organized. Make your answer as specific as possible, and write only what you are asked to address. Also ensure that you structure the body of the essay in the same order you indicated in your introduction. Consider word location and use signals and transitions to direct the reader through your points (see Chapter 3 and 4).

Try to give specific examples and details where possible. Unless you are explicitly asked to state your opinion in the question, do not write about your opinion or how you "feel". The instructor will be evaluating your understanding of the material, not your opinion or feelings.

Lend your essay a clear sense of closure. That is, end your essay with one or two sentences that summarize the main points of your answer. This can be achieved by restating your central idea and indicating why it is important.

Note: Because you will usually be given more time for these types of questions, your instructor will probably expect higher quality in your writing.

 Example 11-7

> *Question: "Describe the theory for the origin of life on Earth and how it is supported by evidence from the following areas:*
> > - *Origin of Earth and its early conditions*
> > - *Fossils, first living things, and conditions for life to flourish*
> > - *Molecular biology"*

Working title: "Theory and Evidence for the origin of Life on Earth"

"The theory for the origin of life is supported by several pieces of evidence, including the origin of the planet, its early conditions and fossil record, and molecular evidence. The Earth's formation is linked to the beginning of the Universe, for which the current theory is the Big Bang. This theory is supported by the Doppler effect because when objects are traveling away from a person, the wavelength of light lengthens, giving objects a redder appearance. Using spectrum analysis, scientists can tell what elements are within stars. When comparing the elements found in our sun to those of other stars, the same elements are observed. However, they are slightly reddened because they are moving away from us, giving them longer wavelengths. This observation suggested that the universe is expanding, which indicates that it had a beginning, as hypothesized in the Big Bang theory.

The theory for the formation of Earth is the dust cloud theory, which states that Earth formed from gaseous clouds that were disk-shaped. All the other things in the solar system formed similarly. Most of the hydrogen and helium in the solar system ended up in the sun, while most of the heavier elements formed the planets including the Earth. In its infancy, Earth had little to no crust and an atmosphere of hydrogen and helium. This stage would have lasted no more than 100 million years. Earth's second atmosphere was composed of the gases volcanoes emitted: CO_2, H_2S, H_2O, ammonia, etc. Then, life began to form—probably in Earth's oceans—and eventually became photosynthetic. Photosynthesis caused oxygen to be produced.

Earth's third atmosphere, high in oxygen levels, supported aerobic life, which in turn evolved to become bigger and more complex. An ideal temperature for life existed because of the internal pressure, heat and radioactive decay, and the friction produced in the initial formation of Earth.

The conditions on the early Earth would have made life itself almost inevitable. The early Earth, which had many volcanic gases as well as lightning, would have formed organic compounds. This has been repeated in experiments today. The oceans would have become an organic soup where protobionts would have eventually formed. Some of the compounds in the "soup" included microspheres and RNA, the latter of which has catalytic as well as informational properties. Furthermore, there was abundant energy and partial isolation, which would have allowed DNA to form. There were replicators, and they had the potential to mutate.

The theory of life is also supported because the first prokaryotes were very simple organisms that could have formed from this "soup". Fossils show that bacteria did exist about 3.8 billion years ago, and that they were extremely simple, but soon began to evolve into more complex forms.

	Molecular biology also supports the theory of life on Earth because it shows that biological molecules have—like organisms—themselves evolved. For example, RNA started out having catalytic functions, but later enzymes began being used for this purpose because they were more efficient.
Body	
Concluding paragraph	All together, these pieces of evidence provide us with information for the formation and origin of Earth and lead to the theory for the origin of life on the planet.

(With permission from Nikolas Franceschi Hofmann)

GUIDELINE:

Proofread your answer.

After you are done writing your answer, read over it and check that all the main ideas have been included. An essay answer usually also tests your writing ability. Thus, be sure to review and proofread your answer for careless mistakes, misspelled or illegible words, and grammatical errors.

GUIDELINE:

Be aware of additional strategies to answer an essay question.

Additional strategies may include:

- Using relevant technical terminology that you learn from your courses to answer the question
- Making relevant and logical connections clear—for example, for cause and effect relationships or drawings with labels
- Making a drawing—this can often assist you in your answer, although your written answer must explain what is in the drawing
- Supporting your answer with evidence and/or examples from class
- Budgeting your time to ensure that you are not forced to rush through your essay
- Leaving time to proofread for grammar, spelling errors, omissions, etc.; too many errors will be credibility killers
- Qualifying your answers when in doubt. It is better to say "about 500" than to say "485" when you cannot remember the exact number
- Avoiding excess information

- Jotting down the remaining main ideas from your outline for potential extra credit in case you run out of time when you are writing an answer
- Writing legibly so your instructor can read what you have written, and you do not lose points unnecessarily

11.4   TIME MANAGEMENT TIPS

GUIDELINE:

Manage your time efficiently.

Be aware of the time, especially if there is more than one essay question you are asked to answer. Decide which question(s) you will answer first. Start with the one you know most about and the one you can answer the quickest.

As a guideline, 2 to 5 min corresponds to 20 to 30 words and will leave you no time for details on a major point; 10 to 15 min corresponds to 50 to 75 words, allowing time for one detail on a major point; 20 to 30 min corresponds to 100 to 175 words and give you time for two details on each major point.

11.5   CHECKLIST

Use the following as a checklist after you have written your answer.

Checklist for Essay Tests

☐   Will the reader get an overview of your answer from reading the first sentence of your answer?
☐   Did you describe the best ideas first?
☐   Did you support your major ideas with examples and facts?
☐   Did you answer the question fully?
☐   Did you stick to the question and did not include extra material?
☐   Did you end with a conclusion?
☐   Did you proofread your essay?

SUMMARY

1. Prepare for an essay exam—study and understand the material.
2. To answer an essay question:
   - Read and understand the question.
   - Jot down notes/brainstorm and make an outline.

- Write the first sentence as an overview sentence.
- Logically organize and present your ideas.
- Proofread your answer.

3. Be aware of additional strategies to answer an essay question.
4. Manage your time efficiently.

PART FOUR

Advanced Scientific Documents and Presentations

CHAPTER 12

Oral Presentations

12.1 ORGANIZATION OF A SCIENTIFIC TALK

GUIDELINES:

Organize your slides.

Know your audience.

The content of a scientific talk is similar to that of a research article with an introduction, results, and discussion/conclusion. Unlike a research article, however, an oral presentation needs to be structured and worded differently from a written document.

Whether you are presenting a short, 10-min conference talk or a full-hour seminar, a good overall format to follow for your talk is the following:

Title slide (optional)

First slide:	**Overview of talk (optional for short talk)**
Next slides:	**Introduction/Background and purpose of study**
Subsequent slides:	**Findings (combined with general approach)**
Final slide:	**Conclusions and main supporting points**
Credit slide (optional)	Acknowledgment of those who worked with you or financed your research.

It is critical that you know your audience and the overall purpose of your talk before you prepare your slides. Depending on your target audience and the type of talk, different emphasis may have to be placed on different sections of the presentation. For example, if you talk is geared toward a non-scientific audience, you may have to include more introductory material and slides, because it will be particularly important to provide sufficient enough background information to avoid losing your audience in the first few minutes. For a scientific audience, on the other hand, introductory slides may not be as critical as more details on results, control experiments, and comparisons and contrasts to findings of others.

GUIDELINE:

Prepare an overview slide for your presentation:
Tell the audience what you are going to talk about.
Talk about it.
Summarize what you have told them.

Start your presentation with an overview by telling the audience what you will speak about: introduce the overall topic, mention how you will present your findings and that you will summarize all before you close. Then move on to provide general background information on your topic.

If you are preparing a longer presentation, you may want to consider using an overview slide at the very beginning to lay out the different parts you will be covering. Follow this slide with an introduction of background material to provide context for your talk.

After the background, explain the purpose for your studies and discuss your results. The main emphasis should be on your key findings. The experimental approach is often only briefly alluded to verbally and in most cases does not require extra slides. Providing an overview of the experimental approach is sufficient; exact details such as amounts of reagents are usually not given in a presentation. However, if the experimental setup is complicated or if you are introducing new or complex methodology you may need to explain the experimental approach in more detail to your audience. In these cases, including additional slides on your approach is absolutely appropriate and necessary.

Most presentations end with a concluding slide, which summarizes and generalizes your main findings. Depending on your audience, a concluding slide may also point out next steps or give credit to colleagues and collaborators.

Shorter talks are usually more difficult to prepare and to present than longer talks. For shorter talks, you need to be extremely selective about what information to present. For such talks, concentrate on the overall, most important information for each of the sections. Generally, this means that you will need to reduce the number of background slides and focus on one or two main findings and their interpretations.

12.2   PREPARING FOR A TALK

GUIDELINES:

Prepare your talk well ahead of time.
Prepare visual aids well ahead.

The two most important points in becoming a good presenter are to be prepared and to practice. Know your audience and their level of expertise. Design your talk accordingly in terms of its direction and necessary background information. Put your slides together well ahead of the talk. Preparing visual aids well ahead of the presentation will give you time to check, replace, or improve them.

GUIDELINES:

Make slides look attractive.

Keep slides simple.

Limit the amount of text.

Visual aids should be comprehensible on their own. They must be clear, legible, and easy-to-understand. They should also be visually pleasing, professional, and free of spelling errors.

Format

For scientific presentations, most of the time you need to use a conference style presentation and layout for your slides, as in a scientific presentation, the intention is to either inform or persuade your audience. Provide sufficient information in text slides as well as with figures and tables to present an overview without an overload of text or figures. Any graphical layout must instantly communicate what the slide is about, must contain relevant details, and must be free of distractions. Note that although slide layouts with large photos and no or only sparse text are favored by some schools of organizations, these are only great for informing and entertaining audiences, such as for keynote addresses, court rooms, and sales meetings—not for scientific conferences.

For the most part, PowerPoint, Apple Keynote, SlideRocket, Google Presentation doc, Adobe Presenter, and so forth can be used to design your slides for a scientific presentation. All of these presentation tools follow the traditional outline, linear slide presentation path. Another interesting new alternative is Prezi, which uses a more free-flow, creative style, nonlinear path to presenting your ideas. Generally, similar zoom-in and nonlinear flow can also be achieved with the more traditional programs mentioned earlier, although information on how to do so is not as readily obvious in these programs.

Colors

To create visually attractive slides, avoid bright colors, nonstandard colors, and red/green (or blue/orange) color contrasts, as some people are color blind to these. Individual slides should also be recognizable as being part of a set (same colors, style of fonts, and emphasis techniques).

Font

Avoid nonstandard fonts, graphs, and abbreviations. Ensure that the minimum font size is larger than or equal to 18 points so lettering can be read by the audience during the presentation. Use a sans serif font type such as Arial—it is easier to read than serif font, such as Times New Roman. Look for high contrast between background and writing or figures. Choose dark text against a light background or vice versa. Medium to dark blue background with white or yellow writing, for example, is commonly used and easy to read. If you need to emphasize text, put the most important information in a larger print size or into italics, or use different colors.

Text

Not only do your slides need to be visually attractive, they also should contain informative headings and cover all key points, yet be simple. Most people can only absorb a limited amount of information at once. Therefore, to ensure that you are not losing your audience, do not clutter your slides with too much text. Write clear but brief bullet points to provide an overview and fill in the rest with your voice. If your slides contain too much text, your audience will concentrate on reading the text and not listen to what you have to say, or they will be listening to you and not pay attention to what is on the slide. Neither case is what you as the presenter really desires.

As a rough guideline, use no more than

- 40 words per slide
- 40 characters per line
- 12–15 lines per slide
- three to five words per bullet point
- six to eight bullet points per slide

In an average presentation, you will spend about 1–2 min on each slide and will have to cover all the information that is contained in it.

Following is an example of a slide that contains too much text.

Example 12-1   **Text slide**

Overview of the Yale School of Medicine

- Founded in 1810, the Yale School of Medicine is a world-renowned center for biomedical research, education and advanced health care.
- The School is viewed internationally as a leader in biological and medical research.
- The Yale School of Medicine has over 900 faculty members and consists of 9 basic science departments and 17 clinical departments.
- The School of Medicine consistently ranks among the handful of leading recipients of research funding from the National Institutes of Health and other organizations supporting the biomedical sciences

To improve a text-heavy slide such as the one shown in Example 12–1, you need to decide which bullet points can be omitted and what information in the remaining bullet points is really important. Then list the important information by itself in a bullet point and omit the rest. A possible revision of Example 12–1 follows:

**Revised
Example 12-1a**   **Text slide**

Yale School of Medicine Overview

- Founded in 1810
- Leader in biomedical research
- Over 900 faculty members
- 9 basic science departments
- 17 clinical departments
- Top biomedical research funding

In the revised version, the bullet points have been reduced to their main piece of information and are visually distributed better on the slide, all of which is preferred by the audience.

If we further revise Example 12-1a, by adding a different background as well as a picture and using different styled and shaded text for the heading, we get a much more visually attractive slide:

**Revised
Example 12-1b**   **Text slide with background and graphic**

Yale School of Medicine Overview

- Founded in 1810
- Leader in biomedical research
- Over 900 faculty members
- 9 basic science departments
- 17 clinical departments.
- Top biomedical research funding

GUIDELINES:
Give figures and tables a title but no legend or caption.
Think graphically.

On slides, figures and tables should have a title but no legend. Do not simply transfer published figures or tables onto your slides. Their lettering is often too small, they do not have a heading, and more often than not, printed figures and tables contain additional information that is not needed for your presentation. Instead, present only figures or tables with clearly visible axis labels, keys, and a large heading. If needed, reconstruct the figure or axis labels to ensure that they are large enough to be visible from the back of the room.

Know that readers prefer visually well-designed slides rather than text-heavy slides. They also prefer graphs rather than tables or text, particularly bar graphs because their message can be quickly understood.

Ensure that your figures and tables are not cluttered or too busy. For line graphs, display not more than 3–4 curves per graph. Similarly, in bar graphs, keep the number of bars to a minimum. In your figures, symbols and shading must be easy to tell apart. Do not include a figure legend or caption on slides, but provide a key for figures if needed.

An example of a slide containing a graph is shown next. Note that in this slide, the title for the graph is at the same time the title for the slide. It is possible to compose such titles if your slide contains a single figure.

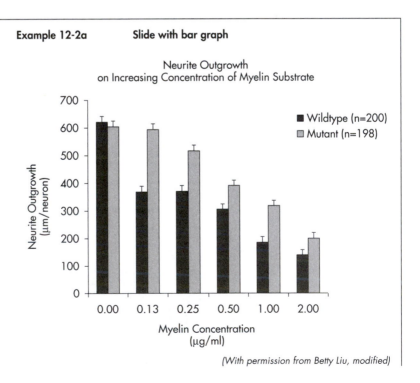

Example 12-2a   Slide with bar graph

Neurite Outgrowth
on Increasing Concentration of Myelin Substrate

(With permission from Betty Liu, modified)

If you have to use a table, aim for a maximum of four columns and seven to eight rows, including the title and column headings. Write all text horizontally. The next example shows a slide containing a well-constructed table.

Example 12-2b   **Slide with table**

Thermal induction of cytokinin overproduction			
		Concentration of cytokinin in medium (nM)	
Mutant	Cytokinin	15°C	25°C
thiA1	iP	0.16	0.23
	[9R]iP	0.09	0.16
OveST25	iP	6.34	60.0
	[9R]iP	0.54	9.51

GUIDELINE:
Prepare notes.

Most people find it helpful to prepare notes for their presentation. You may not need them when you are presenting but might want to use them to practice your talk. For most people, notes also provide the security of knowing what to fall back onto when you are too nervous to remember anything during critical portions of your talk.

To record your notes, use index cards or the speaker's notes field on PowerPoint. Neither will be visible to the audience and allows you to maintain an image of spontaneity. If you decide to use note cards, number them for ease in assembling them.

Use extra-large print for your notes if necessary—they should be easily readable. Write your notes in outline form, not in full sentences. There are a few exceptions when notes should be written out in full:

- The opening of your presentation
- The closing section of your presentation
- Transitions (between slides and sections)
- Quotations (note that there will be few, if any, in scientific talks)

Having the opening sentences available in full will give you comfort in knowing where to find statements when you need them.

To deliver the opening sentences firmly and accurately, memorize them. A good start will give you confidence to continue and allow nervousness to abate. As you conclude your presentation, it is also comforting to know that you have your conclusion written out, especially to ensure that you get the wording correct. Transitions that easily evade you during your talk should also be written out in full to serve as bridges and references.

GUIDELINE:

Practice, practice, practice.

For the best possible presentation, it is of utmost importance that you practice and practice and practice. Without practicing, you will not know whether you stay within the given time limit for your talk. Nor will you know if your flow of words is smooth or if your voice has the right pitch. Practicing will make you realize at least some of these potential problem areas. Consider videotaping yourself, or ask someone else to do this for you. Review the video and note any areas that may need improvement. In addition, take advantage of opportunities to give practice presentations.

12.3   DELIVERY OF A TALK

GUIDELINE:

Arrive early and dress appropriately.

When you have to present a talk, arrive early and appropriately dressed. Familiarize yourself with the surroundings and equipment. Ensure that your presentation file is compatible with the computer connected to the LCD projector, especially if you deliver your presentation on a flash drive or CD. As a back-up, consider bringing your own computer, or send a pdf copy of the file to yourself or the organizer ahead of time. Alternatively, come prepared with a printed out version of your slides, which you may ask to have photocopied and handed out if all else fails.

GUIDELINE:

Use spoken English.

During your talk, use simple words even if you have to apply technical terms. Do not speak the way you write in scientific English. Use spoken language instead.

If you use notes during your presentation, do not read them word for word. Also, do not use the bullet points on your slides as your notes. These bullet points are intended for your audience, not for you as the speaker. Therefore, it is important that you prepare separate notes for yourself.

When you use notes, speak and act normally. Being yourself is the most valuable asset of a speaker. Above all, be positive and enthusiastic about your topic.

GUIDELINES:

Stick to the time limit.

Speak slowly.

It is customary to assign a limit to the length of a presentation. Adhere to this time limit. Nothing angers an audience or the organizers more than a speaker who goes overtime. To help you keep track of time, place a timer on the lectern. On average, you will need one to two minutes per slide. If you find that you need to leave out some important slides to stay within the time limit, do so. Do not speed through your talk. You run the risk of losing your audience if you present your material too fast. Instead, pace yourself—remembering that although you are familiar with the material and have gone over it many times, it is your audience's first time hearing the material, so give them time to digest it. A good talk requires speech that is slower and clearer than normal conversation.

GUIDELINES:

Make sure you can be heard by the entire audience.

Avoid distracting sounds such as "hm" and "uhs".

Do not speak too softly. Soft speech signals that the speaker is uncertain. However, do not blast the audience out of its seat by the volume of your delivery either. In addition, pay attention to the pitch of your voice. Speaking in deeper, fuller tones makes your voice more pleasant to listen to.

Note also that if you use too many distracting sounds such as "hm" and "uh," "ok," and "so," you will appear unprepared and unsure of what will come next. These sounds, if used in excess, can get very distracting for your audience.

GUIDELINES:

Stay within the presenter's triangle.

Keep eye contact.

Be conscious of body movement.

Face the audience.

Use gestures.

The best position for a presenter is on the left or right side of the room or screen (Example 12–3). Either of these positions will ensure that you do not block the projection of your slides and that you do not block the view of the audience.

In any presentation, confidence is the key. The most powerful way to appear confident is to look directly at the audience. Thus, be very conscious

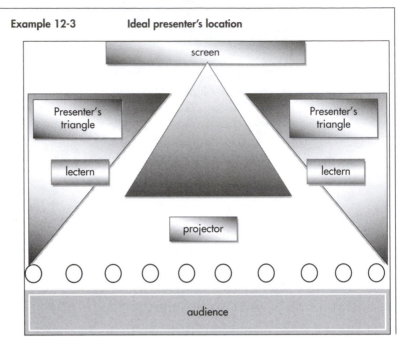

Example 12-3   Ideal presenter's location

of head and eye movements. Eye contact with the audience is essential. The best way to keep your eyes on the audience is by keeping the front of your body facing the audience as much as possible. Look at individuals in the audience, but do not exclude sections of listeners.

Use gestures to reinforce and complement your talk. When you are not using your hands and arms, let them hang naturally, but do not stand rigidly. Use good posture. Do not stick your hands into trouser pockets or fold your arms or hands. Do not use your hands to straighten your clothing, rub your nose, explore your ears, smooth down your hair, or play with keys, bracelets, and so forth.

Your feet should neither be seen nor heard. They should be securely on the floor. Do not teeter back and forth or pace. Stay at the lectern unless you need to point out data on a slide or exhibit. When you show slides, turn halfway toward the screen rather than turning your back on the audience. Face the audience again after pointing out relevant parts on each slide.

If you are using a laser pointer, learn how to use it correctly and sparingly. Above all, learn where the "Off" button is. If your hands tend to shake while you are pointing, consider supporting the pointing hand with your other hand, or resting it on the lectern.

GUIDELINE:

Explain everything on your slides.

Simultaneously show on your slides the information you are providing verbally. Ensure also that you explain everything that is on a slide. Do not skip over information. When you present a figure or table, walk your audience through them in their entirety. For a graph, explain each axis first, and then explain the data and the results they represent. For a table, describe the row and column headings, and then list the key findings in the table field.

GUIDELINE:
Signal the end.

When you conclude your presentation, signal the ending. Say "To conclude, . . ." or "In my final slide . . ." Then summarize your findings and their meaning, briefly discuss the overall significance of your work, indicate next steps if appropriate, and acknowledge those who contributed to the work.

If your words and the tone of your voice do not make it clear that you have finished, just say "Thank you" and stop. If questions are to follow immediately, stay at the lectern. Otherwise, gather your items, and walk back to your seat.

12.4   QUESTION AND ANSWER PERIOD

GUIDELINES:
Ensure that you are in charge.
Stay calm and polite.

Often a scientific presentation is followed by a brief question-and-answer period during which anyone in the audience can ask a question of the presenter. Many, if not most, presenters are nervous about this period, especially about being asked questions to which they might not know the answer.

To deal with difficult questions:

Do

Be courteous in your answers at all times.
Repeat the question to ensure the entire audience has heard it clearly.
Direct the answer to the entire audience.
Admit when you do not know the answer.
Ask the questioner to rephrase the question.
Ask the questioner to talk to you after the session.

Don't

Do not give your audience an opportunity to interrupt your presentation.
Do not maintain eye contact with the questioner when you give your answer.

Do not make up an answer if you do not know.
Do not argue with a questioner.

12.5 USEFUL RESOURCES

For the top 20 tips in using PowerPoint, see Appendix D. For more suggestions and advice on how to create, prepare, and present a scientific oral presentation, consider also the following websites (last accessed October 2012):

- **http://www.catalysis.nl/links/presentations/index.php**—
  explains how to give successful oral and poster presentations
- **http://www.nature.com/scitable/topicpage/oral-presenta-
  tion-structure-13900387**—excellent resource that discusses
  the structure of an oral presentation.
- **http://www2.napier.ac.uk/gus/writing_presenting/
  presentations.html**—useful practical, step-by-step advice on
  how to design an oral presentation
- **https://www.asp.org/education/EffectivePresentations.
  pdf**—a detailed guide to oral presentations
- **http://pubs.acs.org/cen/employment/87/8717employment.
  html**—another site that discusses scientific talks in good detail
- **http://pages.cs.wisc.edu/~markhill/conference-talk.
  html#outline**—provides general oral presentation advice

For useful tutorials of PowerPoint, see (last accessed October 2012):

- **http://www.lynda.com/landing/msofficetutorials.aspx**
- **http://homepage.cs.uri.edu/tutorials/csc101/powerpoint/
  ppt.html**
- **http://office.microsoft.com/en-us/powerpoint-help/create-
  your-first-presentation-RZ001129842.aspx**
- **http://www.actden.com/pp/**

12.6 CHECKLIST

Overall

- ☐ Did you practice your talk several times?
- ☐ Is your talk within the time limit?
- ☐ Are you aware of distracting sounds or habits you show when
  presenting?

Preparation

- ☐ Did you find out who your audience will be?
- ☐ Did you prepare notes?
- ☐ Did you write down
  - ☐ Your opening statement?
  - ☐ Important transitions?
  - ☐ Concluding remarks?

Slides

- ☐ Do you have an overview slide?
- ☐ Is your presentation logically organized?
- ☐ Are your slides informative?
- ☐ Do you have a summary slide?
- ☐ Is each slide logically organized and uncluttered?
- ☐ When preparing your slides, did you concentrate on the main points in each portion of your talk?
- ☐ Do all figures and tables have a title?
- ☐ Did you use visuals where possible rather than text?
- ☐ Is text used sparingly but informatively?
- ☐ Is the font large enough?
- ☐ Are slides/figures/tables kept simple?
- ☐ Are slides attractive? Is color used well?
- ☐ Did you proofread your text?

SUMMARY

ORAL PRESENTATION GUIDELINES:

1. Organize your slides.

Optional:	Title slide
First slide:	**Overview of talk**
Next slides:	**Introduction/Background and purpose of study**
Subsequent slides:	**Findings (combined with general approach)**
Final slide:	**Conclusions and main supporting points**
Optional final slide:	Acknowledgment of those who have worked with you or financed your research.

2. Know your audience.
3. Prepare your talk well ahead of time.
4. Prepare an overview slide for your presentation.
5. Prepare visual aids well ahead.
6. Make slides look attractive.
7. Keep slides simple.
8. Limit the amount of text.
9. Give figures and tables a title but no legend or caption.
10. Think graphically.
11. Prepare notes.
12. Practice, practice, practice.
13. Arrive early and dress appropriately.
14. Use spoken English.
15. Stick to the time limit.
16. Speak slowly.

17. Make sure you can be heard by the entire audience.
18. Avoid distracting sounds such as "hm" and "uhs".
19. Stay within the presenter's triangle.
20. Keep eye contact.
21. Be conscious of body movement.
22. Face the audience.
23. Use gestures.
24. Explain everything on your slides.
25. Signal the end.
26. Ensure that you are in charge.
27. Stay calm and polite.

PROBLEMS

Problem 12-1

Assess the following slide. Explain why this is not a good slide.

TABLE 2

List of Proteins Renatured by the Sparse Matrix Approach and Their Optimum Renaturation Conditions

Protein	Molecular mass (Da)	Structure	Standard assay buffer	Optimum renaturation buffer system	Max activity recovered (percentage of initial value)
BAP	80,000 (29)	Homodimer Zn^{2+}, Mg^{2-} cofactor	100 mM NaCl, 5 mM $MgCl_2$, 100 mM Tris, pH 9.5	0.2 M Na acetate, 0.1 M Tris–HCl, pH 8.5, 30% (w/v) PEG 4000	138
HRP	40,000 (30)	Monomer, heme group, Ca^{2+}, carbohydrate	100 mM CH_3COONa, pH 4.2	0.2 M Mg acetate, 0.1 M Na cacodylate, pH 6, 30% (v/v) 2-methyl-2,4-pentanediol	33
β-gal	540,000 (31)	Tetramer, with independent active sites (32)	Z-buffer (see Materials and Methods)	30% (v/v) PEG 1500	81
Lysozyme	14,388 (33)	Monomer	0.1 M K phosphate buffer, pH 7.0	0.2 M Mg acetate, 0.1 M Na cacodylate, pH 6, 30% (v/v) 2-methyl-2,4-pentanediol	333
Sperm bindin	24,000 (23)	Unknown	Seawater	0.2 M Na citrate, 0.1 M Na Hepes, pH 7.5, 30% (v/v) 2-methyl-2,4-pentanediol	100
Recombinant bindin	24,000 (34)	Unknown	Seawater	0.1 M Na Hepes, pH 7.5, 1.6 M Na, K phosphate	100
Trypsin	23,800 (35)	Monomer (α) dimer (β)	10 mM Tris, pH 8.0	0.2 M Mg chloride, 0.1 M Na Hepes, pH 7.5, 30% (w/v) PEG 400	11
Acetylcholinesterase	260,000 (36)	Aggregates, monomer	0.2 M Na phosphate buffer, pH 7.0	0.1 M Na Hepes, pH 7.5, 0.8 M Na phosphate, 0.8 M K phosphate	9

(With permission from Elsevier)

Problem 12-2

Assess the following slide. Explain why this is not a good slide.

Egg Agglutination Assay

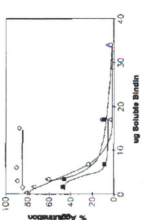

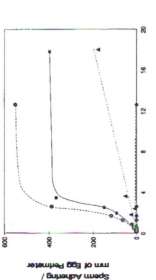

FIG. 4. Species specificity of sperm adhesion in *S. franciscanus* and *S. purpuratus* gametes. The number of adherent sperm was scored as a function of sperm concentration for all possible combinations of *S. franciscanus* and *S. purpuratus* gametes. ●, *S. purpuratus* sperm × *S. purpuratus* eggs; ◆, *S. purpuratus* sperm × *S. franciscanus* eggs; ▲, *S. franciscanus* sperm × *S. purpuratus* eggs; ■, *S. franciscanus* sperm × *S. franciscanus* eggs. Significant numbers of *S. franciscanus* sperm adhere to the surface of *S. purpuratus* eggs. The number of adherent sperm is normalized to account for the larger surface area of *S. franciscanus* eggs.

FIG. 2. Inhibition of egg agglutination by soluble bindin. *S. purpuratus* soluble sperm bindin inhibits egg agglutination non-species specifically. (○) Soluble *Strongylocentrotus purpuratus* sperm bindin added to *S. purpuratus* eggs in the presence of particulate *S. purpuratus* bindin; (△) soluble *S. purpuratus* sperm bindin added to *S. purpuratus* eggs in the presence of particulate *S. franciscanus* bindin; (■) soluble *S. purpuratus* sperm bindin added to *S. franciscanus* eggs in the presence of particulate *S. franciscanus* bindin; (◇) soluble synthetic peptide corresponding to residues 69–130 of *S. purpuratus* bindin added to *S. purpuratus* eggs in the presence of particulate *S. purpuratus* bindin (see text).

(With permission from Elsevier)

Problem 12-3

Explain why the following statements are not good choices for an oral presentation:

1. "Thank you for listening to my talk. I hope it was not too confusing or boring."
2. "On this slide, please focus only on part D, and ignore parts A, B, and C."
3. "This finding is in agreement with that of a previously published result reported by Lopez et al. in 2001, where it was shown that emergence of seedlings is temperature- and humidity-dependent."
4. "It was determined that frogs can hibernate under water for up to six months."

CHAPTER 13

Posters

13.1 CONTENT AND ORGANIZATION OF A POSTER

GUIDELINES:

Design the poster around your research question. Include:

> Title
> Abstract
> Introduction
> Materials and Methods
> Results
> Conclusions
> References
> (Acknowledgments)

Find visual ways to show your work—let the illustrations tell the story.

Focus only on the main points in each section.

A poster is a visual presentation of your work that can serve as a concise communication tool for small groups of people. A poster attracts viewers, serves as an advertisement, and can provide an overview of your work in your absence.

Overall, posters follow the standard scientific format in that they contain all the sections found in a research paper except for the Discussion. Posters include: Title, Abstract, Introduction, Materials and Methods, Results, Conclusion, and References, and optional are Acknowledgments. However, posters do not contain all the details and information found in a manuscript. That is, their word count is much less. Instead, poster sections concentrate only on the main points of each section and present these mostly visually. Think of each part of the poster as a slide that you would show to an audience. As done for a slide presentation, try to keep text to a minimum by using key words and phrases throughout the poster.

- Design the poster around your research question. During your poster presentation time you can use the discussion time with individuals to expand on issues surrounding that central theme.
- Provide an explicit, take-home message. Summarize implications and conclusions briefly, and gear them toward your audience.

Posters are typically presented to small groups of people interested in hearing more about your topic and work. The clearest poster presentation stems from the proper arrangement of information and simplicity of design. Because a poster is a visual presentation, you need to find visual ways to show your work. Use schematic diagrams, arrows, and other strategies to direct the visual attention of the viewer rather than explaining it all in the text alone.

Layout and Design

GUIDELINE:

Aim for about 20% text, 40% graphics, and 40% empty space.

Effective poster design is important. One of the first things you should do when you find out that you have to present a poster is to find out how much space you are allowed. Then, decide on your poster layout.

Sample horizontal poster layouts are shown in Figures 13.1a–13.1c, but know that poster layouts and designs can vary widely. Some have a

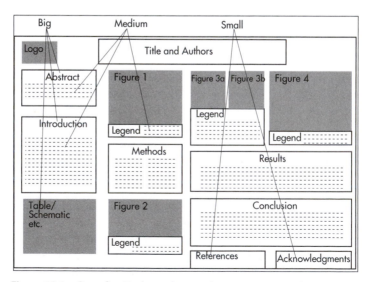

Figure 13.1a Sample - Horizontal poster layouts, asymmetric.

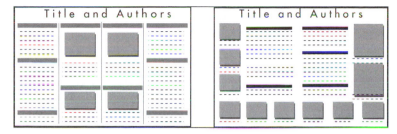

Figure 13.1b Additional horizontal sample poster layouts.

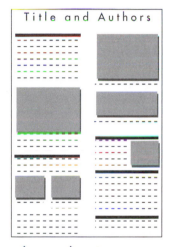

Figure 13.1c Vertical sample poster layout.

symmetrical layout while others do not. Some are horizontally laid out, other are vertical posters.

Although layouts can vary, viewers expect to find certain sections in specific places. The most important text sections (Abstract and Conclusion) are usually placed in the top left and bottom right corner, respectively, as readers read from left to right and top to bottom. The least desirable real estate on a poster is usually on the very bottom. Often, references and acknowledgments and logos are placed there. For clarity

- Present the information in a sequence that is easy to follow. Consider numbering panels to help sequence sections of the poster.
- Arrange the material into columns—most horizontal posters allow for three to four columns, vertical posters usually contain two columns (see Figures 13.1a–13.1c, 13.2, and 13.3.)
- For maximum visual layout and impact, aim for about 20% text, 40% graphics, and 40% empty space. Note that many posters contain much more text. Some even seem to display the pages from entire articles. Such posters are generally considered unattractive by viewers because most people do not want to read through all this text to get to the bottom line.
- Maintain a consistent style.

GUIDELINE:

Design backgrounds well.

The background design is important in the presentation of your data. For well-designed poster backgrounds, use white or muted colors. These colors are easiest to view for a long time and offer the best contrast for text, graphic, and photographs. Do not use more than two or three colors in the background, and do not use bright colors or background designs. All of these can be overbearing and distractive for your audience. Background design or photo background can make distinguishing and reading both text and figures very difficult.

GUIDELINES:

Minimize text.

Make fonts large enough to read.

Because people are attracted to posters that have good graphics, a clear title, and few words, your main objective in preparing text for a poster presentation is to edit it down to very concise language. Use a font size that is large enough to be read from 6 feet away. As a general guideline:

> Titles: 90 point, boldface
> Subtitles: 72 point
> Section headings: 32–36 point
> Body text of main sections: 28–36 point
> Other text (Figure legends, references): 22–28 point

Use bullets and numbers to break text visually and make it more readily available. Stick with the same font type throughout.

13.2   PREPARING A POSTER

Posters can easily be designed using computer programs such as InDesign, LaTeX, Illustrator, CorelDRAW, PowerPoint, and more. Free templates for posters can be found online (see also Section 13.5).

Title

The title should be succinct, clear, and complete (contain all important key terms) and should grab the attention of your audience (see also Chapter 6, Section 6.4). It should be no longer than two lines. The font size needs to be big enough to be read about 10 feet away (1.5-in. minimum). The title is normally followed by the names of the authors and their affiliation.

Abstract

Some posters include an abstract. Others do not. If you are including an abstract on your poster, keep the abstract to a minimum. Include no more than 50 to 100 words. A good example of a possible Abstract follows.

Example 13-1a

ABSTRACT
To examine the food requirements of cardinals, we have assessed the feeding frequency as well as the amount and type of food taken at feeders. We found that cardinals require on average 20 meals of 25 g a day during spring, summer, and fall. Their preferred food source consists of black sunflower seeds. During winter, birds needed about 33% more food per day. When feeding chicks, males and females both required 55% more seeds. Thus, it is important to offer more food to cardinals during winter and when they are rearing chicks.

What you should try to avoid using as an Abstract on your poster is a re-print of what you submitted to the conference committee, such as in the next example. No one in your audience wants to read over this entire, text-heavy Abstract, and the information contained therein can easily be shown under different and more visually appealing headings, such as the Introduction, Results, and Conclusions.

Example 13-1b

ABSTRACT
Isolation and Classification of Developmental
Physcomitrella patens Mutants

Mutants of *P. patens* are useful for the isolation of auxin reg-ulated and signalling genes and can easily be generated by treating either spores or protonemal tissue with UV light. We have collected several UV-induced phenotypical mutants of the thiamine auxotroph ThiA1 of *P. patens* after UV expo-sure. These mutants were screened repeatedly for any auxin effects by growing them on media containing different auxin concentration (0–2 µM NAA). So far we have identified six different classes of mutants:

CLASS A: bud⁻; these mutants do not produce buds in the first 3 weeks of growth under standard condi-tions, but are rescued by auxin in respect to bud and gametophore production (col T-42, spa T-33, spa T-30, col T-73, col T-83)

CLASS B: bud⁻; these mutants never produce buds or gametophores and are not rescued by auxin addition at concentrations of 0.025 to 2 µM for at least the first 3 weeks (spa T-8, col T-78)

CLASS C: gametophore growth not inhibited by auxin (up to 1 µM) (col T-70)

CLASS E: not inhibited by auxin in length and number of filaments (spa T-8, spa T-33, col T-42, spa T-30?, col T-73?)

CLASS F: abnormal branching (spacing and rate) of cau-lonema; rescued by auxin (col T-9, col T-40)

The most interesting mutants so far appear to be the bud-mutants described in class A and B. Auxin in the cell has two possible origins: it is either made by the cell and then relo-cated, or transported into the cell via a putative auxin receptor, after it was produced by a different cell. Auxin within the cell is normally present as conjugates with other molecules and

> is converted to the active IAA form when needed. Both Class A and B mutants were defect in the production of buds/gametophores. Since Class A mutants can be rescued by auxin, they cannot be blocked in the receptor or the conjugate or effect pathways, but must be blocked (at least partially) in auxin production and/or signaling genes. Class B mutants, on the other hand, are not rescued by auxin and can therefore be blocked at several positions: transport, conjugation, effect pathways, or signaling genes.
>
> The long-range goal of this project is to be able to determine which genes of *P. patens* are regulated by auxin. Using the mutants of class A, we would like to determine the time at which the optimal amount of auxin, when added to the growing culture, will induce bud formation. We plan to assess the kinetics of bud formation as well.

Introduction

Use the minimum of background information and definitions to get your viewers interested in the topic (no more than 200 words). Omit unnecessary details. During the poster session, you will be there to fill in details if needed. As most posters are displayed long after a meeting, the information on them must also be self-explanatory.

Ensure that your introduction contains all necessary elements (background, unknown/problem, question/purpose, experimental approach; see also Chapter 6, Section 6.6). These elements should also be signaled clearly.

Example 13-2

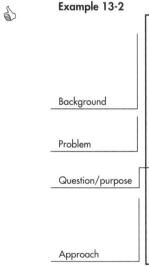

INTRODUCTION
The landscape of the Middle East has been altered by human activity for most of the Holocene period. The rate of these modifications has accelerated in the last century, and today rapid population growth, political conflict, and water scarcity are common throughout the area. All of these factors increase the region's vulnerability to potentially negative impacts of climate change while decreasing the likelihood of successfully emerging region-wide adaptation strategies. In this study, we analyzed climate change in the Middle East during the 21st century as predicted by 18 Global Climate Models. The simulations were run as part of the Intergovernmental Panel on Climate Change Fourth Assessment Report (IPCC AR4) and used the Special Report on Emission Scenarios (SRES) A2 emission scenario, which is the scenario closest to a "business as usual" scenario in the SRES family.

(With permission from Roland Geerken, modified)

Materials and Methods

Briefly describe experimental equipment and methods, but do not include as many details as you would in a research paper unless what you are presenting is a novel or unusual method. Aim for less than about 200 words. Consider using a flow chart to summarize experimental procedures if possible (Example 13–3b) because these are much more visually appealing and quicker to grasp than describing the same experimental procedure in words (Example 13–3a).

Example 13-3a

> **UV mutagenesis of *Physcomitrella patens* to generate auxin mutants**
> To isolate auxin regulated and signaling genes, our laboratory has generated UV-induced mutants by exposing 7-day-old *Physcomitrella patens* moss tissue, which had been blended for 2 min in a commercial blender after dilution of 1:100 (W/v), to 30 s of UV light. After UV light exposure, we prepared protoplasts as described in [2]. Protoplasts were replated and grown for 5–7 days under standard conditions on medium containing 0–2 µM NAA. Phenotypic mutants were selected visually based on abnormal branching or bud formation. Subsequently, mutants were screened as described in [3] and classified as reported previously [4].

Example 13-3b

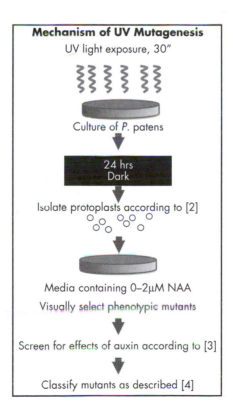

Results

This section is usually the largest portion of the poster. Here you should describe your most important, overall results. Most, if not all, of your findings should be presented in the form of figures and tables. For ease of reading and comprehension for your viewers, ensure a consistent order between your results and the conclusions.

 Example 13-4

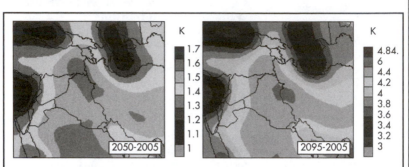

Figure 3 Mean change in annual temperature and precipitation. There is high agreement amongst the Global Climate Models for the predicted temperature change and significant disagreement for the predicted precipitation change

(With permission from Roland Geerken, modified)

Conclusions

Usually, conclusions are very brief, because the poster is indicative of only a portion of the research. Concentrate on your main findings and their interpretation. Mention only two to four main points in your conclusion or summary. If written as bullet points, these findings will be more visually pleasing than a whole paragraph of text.

 Example 13-5

CONCLUSIONS
- Mean annual temperatures will increase by ~4 K by the late 21st century.

- Changes in precipitation are more variable; the largest change, however, is a precipitation decrease that occurs over an area covering the Eastern Mediterranean, Turkey, Syria, Northern Iraq, Northeastern Iran, and the Caucuses.

- Changes in precipitation will have a significant impact on fresh water resources.

(With permission from Roland Geerken, modified)

References

References should be included if the technique is someone else's, but keep references to a minimum and keep them brief. Usually, the use of names, dates, and journal information is enough for the Reference List.

Acknowledgments

If applicable, thank individuals for *specific* contributions to the project, and mention any grant sources (i.e., summer research stipend, faculty grant, etc.).

Figures and Tables

GUIDELINE:

Provide a title and legend for each figure and table.

Include all the figures, graphs, and tables that you will want to point to when discussing your work with someone. This material in a poster should be self-explanatory. Thus, each figure and table should include a legend and a title.

Graphic material also should be simpler than material in published papers. If you have a choice, choose graphs rather than tables—they are more interesting for readers. If you use tables, make them very simple. Use contrast and colors for emphasis.

Printing

Once created, posters can be printed out from commercial printers. If your department or school does not have a poster printer, you can send your file to an online company that prints posters and ask them to mail it to you. Before printing as a poster, however, create a pdf file and print a miniature version of your poster on letter-sized paper to check layout, colors, font size, and design.

Posters for a Conference

If you are planning to present a poster at a conference, you typically have to submit an abstract first. This abstract will be reviewed by a committee, and only if accepted can you show your poster. If your abstract is accepted, review the poster guidelines before starting to make the poster.

Do not use the submitted abstract on your poster. Depending on your discipline, poster abstracts are usually not the same as conference abstracts. Poster abstracts are usually much shorter and less detailed than conference abstracts and, ideally, do not contain more than 50–100 words.

13.3   PRESENTING A POSTER

GUIDELINE:

Be prepared to tell viewers about your work and to answer questions.

Although the material you are presenting should convey the essence of your message, make sure you are at your poster during your assigned presentation time to be available for discussion. In addition, ensure that you have prepared a 5- to 10-min talk, highlighting the key points of your poster (see Chapter 12 for how to structure a talk). Practice this talk in front of your colleagues or professor. Your task as the presenter is also to answer questions as they arise.

During the actual presentation, focus on your graphics. Use your poster as a visual aid but do not read it. Tell viewers the context of your research problem and why it is important, the objectives and how you achieved them, as well as the data and its significance.

13.4   USEFUL RESOURCES

For the top 20 tips in using PowerPoint, see Appendix D. For more suggestions and advice on how to create a poster, see also the following web sites (last accessed October 2011):

- **http://www.swarthmore.edu/NatSci/cpurrin1/posteradvice. htm**—provides excellent advice on designing scientific posters written from a student point of view
- **http://www.ncsu.edu/project/posters/NewSite/Create PosterLayout.html**—gives detailed overview of planning, layout, editing and software
- **http://my.aspb.org/members/group_content_view.asp?group= 72494&id=100256&CFID=1271298&CFTOKEN=27289410**— contains a step-by-step guide on creating a poster
- **http://posters4research.com/templates.php**—great site for free templates

13.5   COMPLETE SAMPLE POSTER

Example 13-6 shows a well designed poster with a short Introduction and Materials and Methods section, clearly laid-out Results section with plenty of visuals, and a short, bulleted Conclusion.

Example 13-6

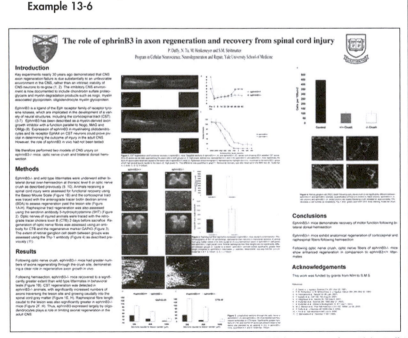

(With permission from Philip Duffy)

13.6   CHECKLIST FOR A POSTER

Use the following checklist to ensure that you have addressed all important elements for a poster:

☐   1. Do the illustrations tell the story?
☐   2. Is the purpose of the research or topic stated precisely?
☐   3. Is your poster abstract—if included—very short and different from your conference abstract?
☐   4. Does the Introduction have the following components?
       Background
       Problem or unknown
       Purpose/topic or review
       Overview of content
☐   5. Did you concentrate on the main points in each section?
☐   6. Is the flow of the panels self-evident to the viewers?
☐   7. Is the topic summarized and interpreted in the Conclusion section?
☐   8. Do all figures and tables have a title and a legend?
☐   9. Is your poster layout uncluttered?
☐   10. Did you use visuals where possible rather than text?
☐   11. Did you keep text to a minimum?
☐   12. Is text written in sans serif font, and is the font large enough?
☐   13. Are exhibits kept simple?
☐   14. Are exhibits attractive? Is color used well?
☐   15. Did you use active voice in the text?
☐   16. Have all jargon and redundancies been omitted?
☐   17. Did you proofread your text?

SUMMARY

POSTER GUIDELINES

1. Design the poster around your research question. Include
   - **Title:** the title must be large enough to be read 6 ft away.
   - **Abstract:** optional; <50-100 words if you use one.
   - **Introduction:** should be self-explanatory and <200 words.
   - **Materials and Methods:** short without much details.
   - **Results:** largest section with many visuals.
   - **Conclusion:** concentrate on your main findings and their meaning
   - **(References):** optional
   - **(Acknowledgments):** optional
2. Find visual ways to show your work—let the illustrations tell the story.
3. Focus only on the main points in each section.
4. Aim for about 20% text, 40% graphics, and 40% empty space.
5. Design backgrounds well.

6. Minimize text.
7. Make fonts large enough to read.
8. Provide a title and legend for each figure and table.
9. Be prepared to tell viewers about your work and to answer questions.

PROBLEMS

Problem 13-1   Poster

Create an attractive poster from one of the following:

1. A set of related laboratory experiments.
2. A paper you have selected from an academic journal.

Problem 13-2   Poster

Would the following be an engaging panel on a poster? Why or why not?

Conclusion

Here we report that recombinant bindin retains the ability to agglutinate eggs species-specifically. We have produced a series of amino- and carboxyl-terminal deletion analogs of *S. purpuratus* bindin to investigate the minimum structure necessary for agglutination. We examined the species specificity of the resulting mutant proteins in egg agglutination assays. Results of these investigations indicate that either the amino- or carboxyl-terminal regions flanking the conserved central domain is sufficient for species specificity. Both ends contain repeated sequence elements that are different in bindins from different species.

CHAPTER 14

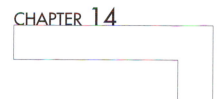

Research Proposals

14.1 GENERAL

To be successful, scientists require funds and have to find sources for this money. These funds are usually garnered through successful grant applications. Grant writing is therefore as central to a scientific career as is writing papers and doing research at the bench.

Before composing a proposal, you must research a funder's priorities. Your proposal must establish a connection between your project's goals and the agency's philanthropic interests. The proposal must succinctly but thoroughly present the need or problem, the proposed solution, and your organization's qualifications for implementing that solution. It must be drafted with much care to boost your chances of success.

14.2 CONTENT AND ORGANIZATION

GUIDELINE:

Follow instructions EXACLTY.

If you have instructions for your research proposal, follow them EXACTLY. Sometimes, proposals receive a bad evaluation grade simply because they do not conform to the required format specified by the instructions. There are many funders, from federal to private, and most have their own ideas on how they like to have proposals structured. There are even more people seeking funding. Your competition will likely be tough. To maximize your chances of obtaining funding, it is therefore important to present information as requested.

If instructions are not provided, the following structural elements offer a starting place:

>   Abstract
>   Background, including Statement of Need
>   Research Design
>   Significance/Impact

Generally, your proposal should answer the following questions:

>   Why this project?
>   Why you?
>   Why at your institution?
>   Why this sponsor?
>   Why now?

14.3   INDIVIDUAL SECTIONS OF A RESEARCH PROPOSAL

GUIDELINE:

Include:

>   Abstract
>   Background, including Statement of Need
>   Research Design
>   Significance/Impact

Abstract

The abstract serves as the executive summary for the proposal. It articulates the highlights from each section. It includes all of the main information covered in the proposal (background, unknown or problem/statement of need, overall objective, general strategy, and significance/impact) in a single paragraph. The Abstract must be written such that it can stand on its own, without the detailed narrative. It must be concise, informative, and complete.

Pay special attention to power positions in your abstract. Put the most effort into writing the first sentence and the overall objective. The abstract in Example 14–1 contains all these elements and starts with a strong first sentence. It is logically constructed and presents the individual elements in the order most reviewers expect to find them:

Example 14-1	Proposal abstract with integrated specific aims
Background	Plate tectonics distinguishes Earth from other terrestrial planets. Plate tectonics arise due to the convection in the mantle of the Earth; and for plate tectonics to occur, some mechanism must exist to compensate temperature-dependent viscosity. However, it is not well understood what this mechanism is.
Problem	In this proposal, we aim **to approach this long-standing mystery through employing a MCMC algorithm that will (1) systematically explore various sampling strategies and (2) seek the fastest forward model calculation by benchmark-**
Objective and Specific Aims	**ing competing Stokes flow solvers, including the pseudo-compressibility method.** To lay a foundation for realistic 3-D

Strategy

Significance

applications, all computations will be conducted in the 2-D formulation. Microsoft Windows OS will be the main platform for developing the Monte Carlo code and the flow solver as well as for analyzing and visualizing the results of MCMC simulations. Stokes flow calculations will be done with the PI's ABC server. Understanding the physics of plate-tectonic convection in the Earth's mantle will have profound implications for our understanding of the habitability of a terrestrial planet and the evolution of life.

(Jun Korenaga, proposal to private foundation; modified)

Statement of need

The statement of need is an essential element of any proposal and must convince the reviewers that there is an important need that can be met by your project.

Example 14-2   **Statement of need**

> Salamander limb deformities detected in California are of concern to public health experts. It is important to explore the underlying causes of these deformities to establish preventive measures. An urgently needed next step in the protection of public health is the examination of the X hypothesis. Investigations of this hypothesis may not only determine the causative agent of the observed limb deformities but also forgo the danger of potential developmental problems in humans.

Objective and specific aims

GUIDELINE:
Clearly identify the overall objective.

Within the Abstract (or directly following it) clearly identify the overall objective of your proposal so that the readers understand what your long-term goal or mission is to accomplish your proposed work. The objective is the most important statement of your proposal. It should follow logically from the statement of need or the problem. Ensure that reviewers can immediately identify this statement. Either place it in italics or boldface or label it, for example, "Objective" or "Goal."

The overall objective can be subdivided into specific (short-term) aims, which can be listed within the abstract with the objective or as a separate section.

Background

Follow the abstract/overview paragraph with background information, which may also include important preliminary results. The Background section describes what is already known about the problem including previous investigators' results and current hypothesis. This section has to convince reviewers that specific aims, once achieved, will have significant impact on the topic in question. Do not just state that you aim to gain scientific knowledge in your field but rather tie the proposal in to some broader scientific or clinical picture.

Generally, readers expect the parts of the Background section to be arranged in a standard structure:

Background:
    Within subsections:
        Background/known
        Unknown/problem/need
        Aim/hypothesis (optional)
    Summary (optional)

The following example starts with general background to provide context and then lists the problem and offers the hint of a possible solution by generally reporting on some preliminary results:

Example 14-3	Background funneling to unknown
Context/Background	Global warming is arguably one of the most pressing concerns of our time. It has been linked to the rapidity of observed climate change—the fact that the Earth's temperature rose by approximately 0.7°C over the last century (the most dramatic increase documented in historic times) and the attendant threat posed by melting polar icecaps, rising sea levels, and potentially, more severe weather patterns. We do not know yet what proportion of this global warming is due to human activity and what is due to natural variations. More important, we lack an effective model to predict precisely by how much the temperature will rise as a consequence of the increase of the levels of CO_2 and other greenhouse gases in the atmosphere of the Earth. In this proposal we aim to . . .
Specific Problem/Need	
Specific aim	

Research Design

The research design section is also sometimes referred to as the Experimental Approach, Methodology, or Strategy. The main function of this section is to propose in detail what approach you will take to address each of your Specific Aims over the course of the proposed project. This section, which typically is also the longest section of a proposal, will receive the most scrutiny during review, because it is the heart of your proposal. You need to present a clear, logical, and achievable solution to the stated need.

To delineate your plans convincingly, the section should elucidate how you have arrived at your findings, what you expect to add to the topic, the specific experimental approaches, the expected outcomes of your experiments, potential alternate approaches, and the implications your project will have.

Organize the section into subsections according to your Specific Aims. Within the subsections, cover:

> *Heading:* **Specific aim/objective**
> Rationale/hypothesis (not always required)
> **Experimental design**
> **Analysis**
> **Expected results (outcomes and significance)**
> Alternative strategies (not always required)

Example 14-4   **Research Design and Methods Section**

I. Laboratory and Field Behavior Study
Objective: To find natural insect predators of the Monarch butterfly, *Danaus plexippus* L., in the California Central Valley and the Sierras.

Approach and Analysis: For our experiment, we will tether live Monarch butterflies to diverse vegetation in various locations throughout the Central Valley and the Sierras to observe what insect predators prey on Monarch butterflies. Vegetation will include various chaparral plants, cacti, as well as trees, including oak, pine, fir, and palm trees. Butterflies will remain tethered to vegetation for 24 hours. Different heights of tethering will be recorded and the exact location and timing will be noted. A total of 1000 butterflies will be used over a 1-month period in the spring at both locations. Using high speed video camera set on the butterflies, any attack on the butterflies will be recorded over the 24-hour period. Data will subsequently be analyzed as to the time before a butterfly was attacked and the type of predator.

Expected Outcome: We expect to obtain information as to what insect predators attack Monarch butterflies and how quickly predators find and attack the butterflies. Findings will give valuable insights into predation dangers butterflies encounter during part of their migration journey through California and provide baseline data to which future findings can be compared.

Impact and Significance

In the last paragraph of your proposal, state the impact and significance of the project. Focus on how human beings will benefit rather than on your own goals. You may also mention evaluation plans such as measurable objectives. The following example shows a typical impact statement.

Example 14-5   Impact and significance

> Novel, bioengineered crops will allow farmers to grow varieties of plants with enhanced productivity, quality, and improved ability to adapt and survive when faced with adverse environmental conditions. This need is particularly urgent in light of the global food crisis, which has put almost a billion people at risk of hunger and malnutrition. By designing new technology and developing varieties that allow more sustainable farming, the project aims to provide innovative and sustainable solutions to the problems faced by crop growers worldwide.

14.4   CHECKLIST

When you have finished writing the proposal (or if you are asked to edit these sections for a colleague), you can use the following checklist to "dissect" the sections systematically:

☐   1. Is the proposed topic original?
☐   2. Are all the components there? To ensure that all necessary components are present, in the margins of the proposal clearly mark the following:
   ☐   Abstract/Overview
   ☐   Introduction/Background
   ☐   Research Design
   ☐   Impact/Significance
☐   3. Is the abstract kept short and within the set limits?
☐   4. Does the abstract contain
   ☐   Context/background (optional)
   ☐   Statement of need
   ☐   Objective
   ☐   Strategy
   ☐   Significance/impact of initiative
☐   5. Is the overall objective stated precisely?
☐   6. Is the unknown/problem clear?
☐   7. Is the approach clearly delineated and does it include a rationale as well as expected outcomes and significance?

SUMMARY

PROPOSAL GUIDELINES
   1. Follow instructions EXACLTY.
   2. Include the following sections in a proposal:
     ■ **Abstract:** Include: background, unknown/problem/need, **overall objective**, general strategy, and significance/impact.
     ■ **Introduction/Background:** Include: background, statement of need, aim/hypothesis (optional).

- **Research Design:** Include: rationale/hypothesis (not always required), experimental design analysis, expected results, alternative strategies (not always required).
- **Significance/Impact:** Write one paragraph, focusing on benefit to field and society.

3. Clearly identify the overall objective.
4. End the proposal with a broad impact statement.

PROBLEMS

Problem 14-1

Write an abstract for a proposal on comprehensive measurements of CO_2 levels in ponds, lakes, rivers, and the ocean in the American Northeast. In this abstract, open with an important sentence or two and then state the problem/need, your overall objective, how you are proposing to solve it specifically, and the impact/significance of the proposed study. Feel free to invent, look up, or search for information on the Internet if needed.

Problem 14-2

What is the problem with the following abstract submitted for a research proposal? Rewrite the abstract such that it would be acceptable for a proposal.

Background: Cardiovirus is a common cause of gastroenteritis. For routine vaccination of Chinese infants, a new cardiovirus vaccine has been recommended.

Objective: To evaluate the impact and cost-effectiveness of the Chinese cardiovirus vaccine program using a dynamic model of cardiovirus transmission.

Expected outcome: Our analysis will indicate the impact and effectiveness of a rotavirus vaccination program.

Significance: Findings can be used to inform policy makers to prevent rotavirus infection.

Problem 14-3

Identify all components of the following proposal abstract/overview (background, need/problem, objective, specific aims, approach, impact):

Abstract

The role that soils play in mediating global biogeochemical processes is a significant area of uncertainty in ecosystem ecology. One of the main reasons for this uncertainty is that we have a limited understanding of belowground microbial community structure and how this structure is linked to soil processes. Building upon established theory in soil microbial ecology and ecosystem ecology, we predict that the structure of belowground microbial communities will be a key driver of carbon and nutrient dynamics in terrestrial ecosystems. We propose to test and develop the established theories by combining state-of-the-art DNA-based techniques for microbial community analysis together with stable isotope tracer techniques. By doing so, we expect to advance our conceptual and practical understanding of the fundamental linkages between soil microbial community structure and ecosystem-level carbon and nutrient dynamics.

(Mark Bradford and Noah Frierer, proposal to private foundation)

Problem 14-4

Point out what the problems are with the following impact statement.

Impact/Significance.

Funding from the Foundation will not only provide for my postdoctoral fellow but also will result in the publication of two papers over the funding period.

Problem 14-5

Evaluate the following short research proposal. Is it clear what the authors are suggesting to do?

Sea urchin gametes serve as a model system in fertilization studies. Their gametes aid in answering important questions in the field and subsequently, in transferring this knowledge to fertilization interactions of even higher organisms such as mammals.

There are several advantages to using urchin gametes. The biological significance of urchin gamete interaction is well established for cell–cell adhesion. Gamete interaction occurs rapidly and synchronously. Gametes are homogeneous populations of single cells and are readily available in large quantities.

The natural habitat of the sea urchin species *Strongylocentrotus purpuratus* and *S. franciscanus* is along the Pacific coast line from Canada to Baja California. *S. purpuratus* are found in the intertidal zone, while *S. franciscanus* normally prefer somewhat deeper water. However, the two species are often intermingled just below the low water mark. They both have overlapping breeding seasons, but fertilization appears to be largely species-specific. Spawning occurs in December through March, depending on the location.

The urchins can be collected when they are ripe to obtain their gametes. When the urchins are injected with KCl, they spawn and eggs can be collected for fertilization and egg agglutination experiments; sperm can be collected for fertilization and sperm adhesion studies as well as for the isolation of the protein bindin, thought to play a critical role in fertilization and species specificity.

During fertilization, species specificity is determined by the following events: The sea urchin egg is coated by a thick jelly layer, which consists largely of sulfated polysaccharides. The egg plasma membrane is also surrounded by a thin layer called the vitellin layer. The sperm have to pass through the jelly layer in order to adhere to the vitellin layer and fertilize the egg. When the sperm come into contact with the jelly layer, the acrosome reaction is triggered. During the acrosome reaction, the membrane of the acrosomal granule fuses with the overlaying sperm plasma membrane. Bindin is exposed after exocytosis of the granule. Bindin adheres to a vitellin envelope receptor, which contains sulfated, fucose-containing polysaccharides. Extension of the acrosomal process occurs, with bindin coating it on the outside. Then, the acrosomal process penetrates the vitellin envelope. Finally, fusion of the plasma membrane of the sperm and egg occurs.

We are particularly interested in the protein bindin and its structure–function relationship. Specifically, we are investigating how bindin adds to the specificity of sea urchin fertilization. Key questions we are interested in answering include: Is bindin the only sperm protein involved in determining species specificity? What is bindin's receptor on the egg surface? What specific portion of the bindin protein is responsible for species-specificity?

Aside from its role in sperm adhesion, bindin also seems to aggregate in larger clusters. These clusters can cause sea urchin eggs to stick together, and this interaction is species-specific. It may be possible to use this fact as an assay for determining activity of the protein.

In summary, sea urchin fertilization allows us to gain important insights into gamete interactions and specificity. Learning more about the protein bindin and its role in this process will poise us to search for similar proteins and interactions in mammalian cells.

CHAPTER 15

Job Applications

15.1   CURRICULA VITAE (CVs) AND RÉSUMÉS

GUIDELINES:

Tailor your CV or résumé specifically to the position and to the organization.

Your CV or résumé should be well presented and flawless.

To enter into a scientific career, you need to be familiar with documents required for job applications. Your *curriculum vitae,* commonly referred to as a CV, or your résumé is typically the first item that a potential employer sees.

Curriculum Vitae

A CV is a summary of your educational and academic backgrounds, which is used primarily in academic and medical settings. CVs are usually very comprehensive and elaborate on professional history including every term of employment, academic credential, publication, contribution, or significant academic achievement. Do not underestimate your skills or overlook those you may have learned and practiced in lab courses. Typically, CVs have no length restrictions, but it is essential that your CV is clear, concise, and honest.

The most effective CVs are those that are customized to a specific application. Therefore, organize and present your information based on the interests of the employer, highlighting your main qualities on the first page (or two). If needed, do your homework and obtain pertinent information on this employer and the position to which you plan to apply.

CVs differ widely, depending on personal preference, where you are applying to, and your experience. The following list will give you an example of what categories you may include in your CV:

Name & address (no personal information)
Education, with degrees & dates (most recent dates first)
Clinical certifications, with dates
Employment history, brief description & dates; most recent dates first
Grant funding
Leadership and service
Laboratory skills
Publications—name in bold, co-first authors easily identified
Invited presentations and seminars
Teaching experience
Professional qualifications
Certifications and accreditations
Computer skills
Language proficiency
Unique technical abilities
Professional memberships
Honors and awards

Following is a sample CV that serves as an example of the appropriate format for a *curriculum vitae*. Note that educational qualifications and work experience are listed in reverse chronological order.

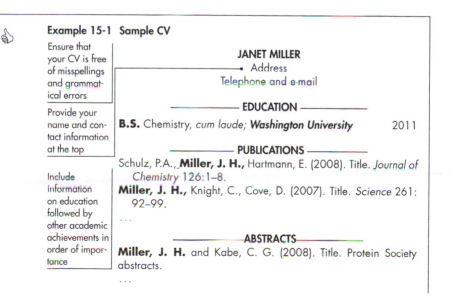

Example 15-1 Sample CV

Ensure that your CV is free of misspellings and grammatical errors

Provide your name and contact information at the top

Include information on education followed by other academic achievements in order of importance

JANET MILLER
Address
Telephone and e-mail

——————— **EDUCATION** ———————

B.S. Chemistry, *cum laude;* ***Washington University***   2011

——————— **PUBLICATIONS** ———————

Schulz, P.A., **Miller, J. H.,** Hartmann, E. (2008). Title. *Journal of Chemistry* 126:1–8.

Miller, J. H., Knight, C., Cove, D. (2007). Title. *Science* 261: 92–99.

. . .

——————— **ABSTRACTS** ———————

Miller, J. H. and Kabe, C. G. (2008). Title. Protein Society abstracts.

. . .

Use formatting to categorize and highlight information	——————— **TALKS**——————— ***Synthesis of Y*** VIIth International Conference on Developmental Biology of the Sea Urchin. Woods Hole, Massachusetts 2007
Ensure perfect alignment and spacing of dates, lists, and subsections	——————— **MEMBERSHIPS**——————— ***American Society for Cell Biology (ASCB)*** ———————**AWARDS AND HONORS**———————
List occurrences in reverse chronological order	***Johns Hopkins University Provost's Award for Excellence in Research*** Fall 2011 ***John Hawkins Award*** for Undergraduate Research May 2010 ***First Place Poster*** - Society for Developmental Biology Conference 2010 ***Phi Beta Kappa*** Spring 2008
	———————**SPECIAL SKILLS**———————
End with References that you may or may not list.	Operation of HPLC and NMR ———————**REFERENCES**——————— Available on request

Résumés

In contrast to a CV, a résumé contains only experience directly relevant to a particular position. Résumés are primarily used for seeking employment in the private sector. They include a summary or listing of your relevant job experiences and education. Your résumé should be tailormade according to the position for which you intend to apply. It should be job-oriented, goal-specific, and very concise. Typically, résumés for industry are much shorter than CVs sent to academia, between 1 and 4 pages, depending on your job experiences. Unlike in a CV, the emphasis of a résumé is not as much on academic achievement and more on professional or work experience. Thus, topics such as memberships in honor societies, talks, abstracts, and honors and awards usually do not appear in a résumé (especially if you have many work experiences to include). However, skills you may have acquired should be highlighted, for example in a separate "Technical Expertise and Skills" section (see the sample résumé following for ideas and wording on what to include in such a section). Although you may be surprised what you can include in such a section based on what you may have learned during your time in college or university, and particularly if you did any undergraduate research, be careful not to overstate your skills unless warranted. Résumés may also contain broader experiences, outside of work, if relevant to the position you are seeking.

Common mistakes in composing a résumé include:

- Not putting your best or most important and relevant experiences near the top

- Listing items in chronological instead of reverse chronological order
- Mentioning specific courses simply because you particularly enjoyed them rather than for their relevance to the job

People who review your application will look at your résumé as well as an accompanying cover letter very carefully. They not only evaluate your experience and skill level but also note the layout, grammar, and spelling of your documents. Creating a CV or resume that is easy to read results from the strategic use of formatting such as bold, italics, spacing (margins and line spacing), and font. Do not put everything in the same type. Rather, format the document such that categories are clearly distinguishable and that your résumé stands out from others. In addition, it is essential that these documents are well-prepared and error-free.

Many employers request a one-page résumé. Although tight, this one-page limit helps streamline the content to only relevant information. You will need to be very selective on what to include and where on the document to place this information. Starting a second page to include a pizza delivery job and a high-school play is not worth the extra page.

Example 15-2

Provide your name and contact information at the very top	**RÉSUMÉ** **Jorge Garcia Colon** **CONTACT ADDRESS** Address; Telephone; Fax; e-mail **PROFILE** • Highly motivated organic chemist with experience in organic synthesis, process chemistry, formulation, and chemical analysis
Ensure that your résumé is free of misspellings and grammatical errors	• Skilled in synthesis of natural products with biological activity, heterocyclic chemistry for cancer, HIV • Excellent team player with good verbal and written communication skills
Place most important and relevant experiences and skills at the top; order remaining sections by importance and relevance	**EDUCATION** 2010 **BS in Chemistry**, Concordia College, Minnesota **TECHNICAL EXPERTISE AND SKILLS**

	Chemistry: Expertise in synthetic organic chemistry; Familiar with operation/data analysis of NMR, HPLC, IR, UV
Use formatting to categorize and highlight information	*Biochemistry:* Experience in fluorescence spectroscopy, circular dichroism (CD), mass spectrometry including MALDI
	Computers: Experience in DOS/Windows and Mac OS systems—graphics and database software using both systems
	Writing: Experience in writing scientific papers
	Speaking: Experience in making presentations to scientific peers

List experience in reverse chronological order	**PROFESSIONAL EXPERIENCE**
	July 2011–present **Chemist, Abbott Laboratories**
	• Responsible for designing and conducting complex, multistep synthesis
Ensure perfect alignment and spacing of dates, lists, and subsections	• Synthesized carcinogenic substances for chemical toxicology research
	June 2010–July 2011 Technical Assistant, ABC University
	• Assisted in research on biology of carcinogens
	• Synthesis of carcinogenic compounds for cancer research

PROFESSIONAL MEMBERSHIPS
Since 2010   American Chemical Society

LANGUAGES
Spanish   native speaker; excellent oral and written Spanish language skills

15.2   COVER LETTERS

GUIDELINE:

Tailor the cover letter specifically to the position.

When you are applying for a job, you will have to send not only a CV or résumé but also a cover letter. Your cover letter creates a professional impression from the outset. Like the CV, it should be flawless, because it can make or break your application. It should also be pleasing to the eye, and thus, well placed on the paper.

In your cover letter, highlight the most relevant parts of your CV that can show your potential employer what you can do for them. Whenever possible, address your cover letter to a specific person, even if that means you have to call the organization to inquire about the name. Check (and double-check) the letter for spelling and grammatical errors. If you are sending a hard copy, print it onto good-quality paper. Preferably, do not send more than a one-page cover letter.

Following is an outline for the general organization of a cover letter:

Opening paragraph
    State the purpose of your letter.
    Mention how you heard about the job.
Middle paragraph(s)
    Highlight your past accomplishments.
    Describe your goals.
    Explain why you believe you are a good fit for the position.
Closing paragraph
    Mention any enclosures (CV, publication samples, Personal Statement, etc.).
    Make positive closing remarks.

Example 15–3 shows a sample cover letter that follows the outline given previously:

Example 15-3 Cover Letter

Date

Use the correct title for the person you are addressing — Dear Prof. Ying:

Purpose and name of position — I am writing to apply for the **Technical Assistant position** as posted on University XXX's web site. I have always been very enthusiastic about working in a laboratory environment and would look forward to the opportunity to assist in the day-to-day routine of a neurobiology lab.

Accomplishments and goals — In May 2011, I will complete my undergraduate studies at the University of California, San Francisco with a BS degree in biology. For more than a year, I have done volunteer work in a research laboratory at the same university, assisting in the preparation of buffers, solutions, and helping in the operation of an HPLC. I have greatly enjoyed this experience and would like to build on and expand my gained knowledge in a laboratory setup by starting my professional career in this environment. It would be an honor to join your research group as a Technical Assistant. I believe that

Reason for fit — my education, talents, and acquired skills as an undergraduate volunteer will be valuable assets for your laboratory.

Enclosures and ending on positive note — Please find enclosed my résumé. The contact information for my references is included, and letters of recommendation will be arriving under separate cover. Thank you for your consideration. I look forward to hearing from you.

Sincerely yours,
John Smith
Enclosures

15.3 PERSONAL STATEMENT

GUIDELINES:

Write an engaging personal statement.

Address the topic or question.

Discuss relevant personal, academic, or research experiences.

Discuss future goals and plans.

Tailor your statement to the particular department and faculty.

Applying for graduate or professional school or post-graduate employment is usually highly competitive. Before you apply, collect as much information as possible on the particular school, program, faculty, or employer you are interested in. Consider speaking to a graduate school advisor or other contact person to obtain further insight.

Oftentimes, committees are sorting through large numbers of applications and need some way to select candidates from the pool of applicants. To aid in the selection of candidates, you may be asked to submit a piece of writing, called a "personal statement," in addition to tests scores and academic transcripts. This written statement provides the admission committee a chance to distinguish you from other applicants. Because your essay can be the deciding factor on whether you are accepted or rejected to an organization, it is essential to take great care in preparing this part of your application.

Your personal statement generally falls into one of two categories:

- A general, comprehensive personal statement, which is usually the type expected for medical or law school applications.
- A response to very specific questions: If an application asks you to answer specific questions, you should respond specifically to the question being asked.

Content and Organization

Regardless of the particular form of the personal statement, committees are interested in learning why they should let you into their organization. To address this concern, you need to answer the following questions:

Why you?

Why this field?

Why this school or employer?

To answer the first question, explain what is special about you or your life story, what your academic interests are, what your career goals are, what your personal characteristics are, what your skills are, and why you are better qualified than others.

For the second question, give specific reasons why you are interested in a particular topic and what specific skills and experiences you would bring to the prospective field of study or employment. Discuss your personal and/or academic background, your research experiences, and why you plan to attend graduate school or work for this particular field.

To answer the third question, give specific reasons why you have chosen this particular school or job to apply to, and state what particularly valuable perspective you will bring to the specific department. What about the department's curriculum structure or general approach to the field make you interested in being a student or employee there? Tailor your essay to match the program to which you are applying.

If you are interested in working with a particular faculty member, consider contacting the faculty member directly. Inquire with the faculty whether there is an opening in the laboratory before you apply to the particular program. Indicate any prior communication with a faculty member about a position in your essay. If you do not want to limit yourself to only just one faculty member, show in your essay how your interests and abilities match the program and department overall.

There are several different ways you can structure a personal statement. The most common format consists of an introduction, a body, and a concluding paragraph.

> *Introduction.* The introduction, especially the first sentence, is the most important part of your statement. Therefore, the first sentence should be attention-grabbing, creative, compelling, and short.

> *The Body.* The body should include roughly three to five paragraphs. Here you should support the main statement made in the introduction. Include your experiences, accomplishments, or any other evidence as well as future goals. You may also want to include a short summary of your educational background.

> *Conclusion.* End your personal statement with a conclusion. For example, explain why you are a good match to the department or faculty.

What Not to Write About

Considering that a personal statement is a professional document that communicates your value as an applicant and potential future addition to a school or company, do not personalize or dramatize the assignment as a chance to tell 'your story'. You should also not write about all your chronological experiences. Rather, select and write about the experiences that make you a valuable applicant.

Similarly, do not discuss about how hard you will work and how badly you want to be 'a doctor'. Committees assume that everyone wants that opportunity. Instead, focus on *why* you want to work so hard toward this goal.

Background Research and Planning

To familiarize yourself with the department to which you are applying, do your homework. Conduct some background research and inquiries. Most importantly, start early. Writing an effective personal statement takes time as you have to consider the topic, plan your points, structure your argument, draft the essay, revise it, and write a final version. Before you begin to write, study very carefully the essay directions on the application materials and then follow them to the point.

Examples of Personal Statements

Following are two examples of personal statement writing toward graduate school:

Example 15-4a   *Essay for graduate school application*

Orli Steinberg   Application to Graduate Program

As a young redheaded child, I had the privilege of being schlepped along with my mother during house visits to her teenage student families. She was a special education teacher who regularly checked in on her students. For my mother, bringing me along aided her in connecting with the families. For me, meeting these diverse people and seeing the different living styles fostered my natural interest and curiosity about the social world. My friends at school called me "The Psychologist" ever since I was eleven years old. I always knew that I would be a psychologist; but this is not why I decided to pursue a Ph.D.

Introduction—describes the applicant's past experiences and character

I had a plan: to learn about human beings living in different cultures around the world and to broaden my horizons. I am half American and half Israeli, and my goal was, once released from the IDF (I know my way around an Uzi), to explore Europe and North America and finally settle in the United States. Working for ELAL Israeli Airlines for ten years made my plan possible. I lived in Amsterdam, Budapest, and Toronto, experiencing life in multicultural cities and interacting with people living in different societies. I loved these interactions; but this is not why I decided to pursue a Ph.D.

While attaining a B.A. degree in Psychology from the International English Speaking Program at ELTE University in Budapest, I took 40 courses in all areas of psychology and psychology research methods and enjoyed every single one. I wrote numerous research-oriented papers and conducted three major studies in the main areas of psychology: social, developmental, and personality. In my final year, I conducted a General Psychology Thesis, which I defended orally in front of an academic committee. At the end of my defense, the committee members applauded me and acknowledged that they were intrigued by my work, enjoyed my presentation, and only rarely meet a student who is so passionate about research. This passion for psychology and psychology research is the reason that I decided to pursue a Ph.D.

Body—discusses the field of interest

Body—this particular applicant has had past work experience in the field and describes these here

I have always been fascinated by human behavior. Although attending a Ph.D. program was not an option for me until now, I focused on jobs that would enable me to learn about the human mind and the social world. Working for ELAL at their global branches, I excelled both as a supervisor and as an instructor and was encouraged to pursue my M.A. in Human Resources Management. Subsequently, I was the global HR manager for an AIDS foundation in Los Angeles. Recently, when asked if I was ready to enter a 5- to 6-year Ph.D. program, my response was that I have been ready for quite some time. Finally, my personal and financial circumstances allow me to continue with my true passion as a scholar, learning and contributing to academic research. A sign of my commitment perhaps is the fact that I voluntarily left a lucrative field, to return to academic research and learning. In pursuing my Ph.D. in Psychology, my ultimate goal is to establish myself as an academic researcher and a psychologist, conducting serious research to expand human knowledge.

Body—further description of past work and research experiences

The research that I conducted already at the B.A. and M.A. levels has prepared me for entering a Ph.D. program that emphasizes excellence in research. My managerial roles at both workplaces have further equipped me with the necessary means to entering a Ph.D. program that prepares its students for leadership. My global explorations, which have catered to my interest in global social phenomena, have prepared me for pursuing multidisciplinary and cross-cultural research.

Body—further description of past work and research experiences

While obtaining my B.A. degree in Psychology, I have revised and developed research tools under tutor supervision. I have conducted research that discussed motivation and examined the effects of negative mood on persistence in problem solving. I have investigated human helping behaviors in emergency and emotional situations, and, endorsed by the Ministry of Education, I have explored the role of media in teenagers' lives in cooperation with a junior high school in my hometown in Israel. In short, I love research and the life of the mind and have spent the last several years preparing myself to pursue my goal as an academic researcher and psychologist.

Applicant addresses the reason for applying the this particular school/ department

I have thoroughly reviewed dozens of psychology programs and came to the conclusion that the psychology graduate program at University AAA not only is one of the finest in North America, but is by far the most suitable for my aptitude, interest, and aspirations. When I explored options for Ph.D. programs, I reviewed also your faculty publications. I am drawn to the research of several of your faculty members, including Professor K. H. for his work in culture and cognition and in culture and self-concept. I am particularly attracted to his idea that psychologists should be at the cores of what cultures really are. I also admire Professor S. L. for her work in the social bases of self and identity and for her intriguing research on transference, as well as Professor D. K. for her work in emotion and social interaction, and her fascinating research of "naïve realism". I would be honored to have either of these faculty members as my mentor.

Conclusion— discusses envisioned, positive outcome if accepted

I am extremely excited about my academic future and I truly believe that University AAA is a great fit for my academic interests and goals. I look forward to the challenge of ultimately exiting your gates as an accomplished researcher who will contribute to the field of psychology and to society as a whole.

(With permission from Orli Steinberg)

Following is a second, well-written essay.

Example 15-4b	***Essay for graduate school application***
Mónica I. Feliú-Mójer	Application to Graduate Program

The first paragraph introduces the applicant

Growing up in a rural area of a small town in northern Puerto Rico, I was surrounded by nature and have since been captivated by its wonders. Collecting stones and insects was my favorite hobby, and my parents gave me a microscope for my ninth birthday. This fueled my scientific interest, making me wonder how nature worked, how everything was formed, and what the biological basis for events that surrounded me was. At 11 years old, when my father was diagnosed with a mental disorder, my interest was guided toward understanding how the brain works.

My high-school academic credentials granted me early admission to the Biology program in XXX University. There, two words would eventually change my life: "Try research." These were the words of my freshman year Biology professor, as she handed me an application for a summer research program at YYY University, one of Puerto Rico's medical schools. There, I had my first research experience as a member of Dr. R.J.C.'s neurophysiology lab. After spending the summer there, my interest and commitment to research were evident and I was selected as part of the MBRS program. During the two years I spent as an undergraduate in Dr. R.J.C.'s lab, I studied the behavioral and functional effects of repeated cocaine administration on rats, focusing on the functional changes in the noradrenergic neurotransmitter system. I had the opportunity to present my projects in several scientific forums and meetings, giving me the unique chance to interact with my peers and become more actively involved in the Neuroscience research community. My undergraduate research experience solidified my interest in neuroscience and my desire to do research, pushing the boundaries of knowledge in this fascinating and still mysterious field.

Since that initial research experience, I have been able to investigate other areas of neuroscience and physiology. In a research internship in Dr. P.A. Lab at ZZZ University, I worked in the cardiovascular physiology field, doing *in vivo* evaluations of a mouse model of congestive heart failure. As a member of Dr. P.A.'s lab, I could compare differences between clinical research and the basic science I had performed earlier on in Dr. R.J.C.'s lab. My experience at ZZZ provided me with the opportunity to explore a completely different field of research and the reassurance that basic science, specifically in neurobiology, is my passion.

The next paragraphs state applicant's areas of interest

To strengthen my skills and gain experience in Neuroscience, I moved to the United States in 2004, to work in Dr. S.M's lab at the Center for Learning. At the age of 20, leaving my parents, friends, and family and giving up my first language was not easy. However, my will and determination to continue advancing in the neuroscience field proved to be strong, making my first experience as a full-time researcher a very enjoyable and productive one. As a Research Assistant in Dr. S.M.'s lab, I have explored the molecular mechanisms of synaptic plasticity using diverse molecular, cellular biology, and biochemical techniques such as Western blot analysis, Yeast two-hybrid assays, dissociated neuron culture and cell culture, recombinant DNA cloning, immunoprecipitation, and generation and analysis of transgenic mice. In addition, I am responsible for administrative duties such as ordering, safety, overseeing laboratory animal issues, and keeping a common stock of drugs for the lab. Recently, one of my projects was published in the *Journal of Neuroscience Methods*. As a member of the S.M. lab, I have gained a richer understanding of

Additional background information/ past work/ research experiences the realities, hardships, and beauties of science, and also of the importance of social sensitivity—the ability to work with people from different countries, cultures and scientific styles. Here, as I strengthen my laboratory skills, I practice how to handle multiple tasks and responsibilities, work hard, commit to goals, and overcome disappointments.

Additional background information and life philosophy

Being born, raised, and educated in Puerto Rico, where scientific opportunities and resources are scarce, along with having Spanish as my first language, has challenged me to become more capable in order to compete and succeed. As a scientist, I understand the importance of making the population aware and informed about the advances being made in Neuroscience, a very prolific, interesting, and fast-growing field. As a minority, I know that specific cultures and ethnic groups have different cultural and linguistic needs. Research, its results, and its impact on specific communities need to be addressed in a sensitive and nonstereotyped way. Being able to overcome these cultural and linguistic barriers gives me the unique opportunity to be a bridge between my field and my community and to encourage the participation of underrepresented minorities.

Reason for wanting to pursue graduate school and explanation of why this particular one

Graduate school is the next step in my pursuit of a career in Neuroscience. H University's Graduate Program in Neuroscience, as well as its comprehensive and multidisciplinary approach, fits my prospective of a graduate program that will expand my knowledge base, grant me access to the most relevant and provocative questions, and allow me to meet mentors and peers who challenge me. With a broad variety of research interests among the faculty and the chance to perform my research in collaboration with various laboratories, H University offers me the opportunity to train in a highly interactive and diverse scientific environment. The high-quality training that H offers me will help me achieve my long-term goal of becoming an independent researcher, performing insightful and cutting edge research, teaching others, and giving people the chance to pursue their passion, as I am doing.

(With permission from Mónica I. Feliú-Mójer)

Candidate Selection

Typically, your personal statement will be read by people who serve on an admissions committee in the department to which you are applying. Carefully consider this audience. Know the areas of specialty of the department so you can gear your essay accordingly. You will want your statement to engage the readers and to set you apart from the rest of the stack. For most people, the big challenge is to find an angle to make their personal statement engaging and interesting. So spend time finding a good "hook" for your statement.

Admission committees look for how well an applicant writes and constructs an essay that is relevant and informative. They also note how well an applicant attends to details and demonstrates critical thinking and abstract reasoning, how well the essay is organized, and if the level of detail is appropriate.

Length, Tone, and Style

Usually, personal statements are between 500 and 1000 words long. Do not exceed any word limit given, and answer all the questions being asked.

The tone of the essay should be balanced or moderate. Do not sound too casual or too formal. Portray confidence and use an active voice. Although personal statements use first person ("I," "we," "my," etc.), avoid overusing "I." Instead, alter between "I" and other first person terms, such as "my" and "me" and use transition words, such as "however" and "in addition."

Make sure to allow time for revision. Ask others to read your essay and to give you critical and honest feedback to help you improve your essay. Revise your essay until you are satisfied with it. Ensure also that a title and your name are included on the first page.

15.4   LETTER OF RECOMMENDATION

GUIDELINES:

In the letter of recommendation, highlight only positive qualities.
Pay particular attention to the first sentence.

Along with your CV, cover letter, and personal statement (if requested), you will probably need one or more letters of recommendation to secure a good position. When you have to ask someone for a letter of recommendation, explain exactly why the letter is needed and how important it is to you. Always offer to provide information that makes the writing task easier (CV, list of accomplishments, publication list, the due date, means of transmission [mail or e-mail], correct address of recipient). Make fulfilling this request as easy as possible for the letter writer. Include any instructions and information on the person or organization that the letter is supposed to go to. Explain to the recommender *why* this opportunity is important to you and *how* it fits into your overall career plan. If the writer cannot or will not provide you with a letter, accept this decline gracefully.

Plan your request. Ask someone who knows you well enough to include details about you as a person in the letter (an important reason as to why it is important to establish good relationships with your professors, supervisors, and peers). The writer should ideally write well, have experience composing letters of recommendation, and have the highest or most relevant job title. As a general rule, request your letter at least a month in advance. Some recommenders who have very busy schedules may even ask you to write a first draft yourself, which they will then use as a basis for their letter of recommendation.

If you have been asked to provide a first draft for a letter of recommendation, avoid a long list of general praises. Instead, write something about your unique personality/traits and about particular situations where you made a difference. Topics you may want to write about include the following:

- Academic performance
- Honors and awards
- Initiative, dedication, integrity, reliability, etc.
- Willingness to follow school policy
- Ability to work with others
- Ability to work independently

15.5   CHECKLIST FOR A JOB APPLICATION

OVERALL

☐   1. Did you tailor your CV or résumé specifically to the position and to the organization?
☐   2. Is your CV well presented and flawless?
☐   3. Did you tailor the cover letter specifically to the position?

PERSONAL STATEMENT

☐   1. Is your statement engaging?
☐   2. Did you address the question or topic?
☐   3. Did you discuss your future goals and plans?
☐   4. Did you discuss how your interests match the department and faculty?
☐   5. Did you show what you can offer (e.g., hard work, research skills)?
☐   6. Did you discuss relevant personal, academic, and research experiences?
☐   7. Did you get feedback on the content?
☐   8. Did you stick to the word limit, i.e. write an essay 1–2 pages in length?
☐   9. Did you use positive language?
☐ 10. Did you proofread your essay?

LETTER OF RECOMMENDATION

☐   1. Are only positive qualities highlighted in the letter of recommendation?
☐   2. Has particular attention been paid to the first sentence?

SUMMARY

JOB APPLICATION GUIDELINES

1. Tailor your CV or résumé specifically to the position and to the organization.
2. Your CV should be well presented and flawless.
3. Tailor the cover letter specifically to the position.

PERSONAL STATEMENT GUIDELINES

1. Write an engaging personal statement.
2. Address the topic or question.

3. Discuss relevant personal, academic, or research experiences.
4. Discuss future goals and plans.
5. Tailor your statement to the particular department and faculty.

LETTER OF RECOMMENDATION GUIDELINES
1. In the letter of recommendation, highlight only positive qualities.
2. Pay particular attention to the first sentence.

Appendix A

COMMONLY CONFUSED AND MISUSED WORDS

Learn to distinguish between words that more casual writers carelessly interchange. This list explains the meaning and use of the most commonly misused words scientific editors encounter. Strunk and White, Fowler, and Perelman have similar lists.

ABILITY, CAPACITY

Ability	The mental or physical power to do something or the skill in doing it. Ability can be measured; capacity cannot. (Some microorganisms have the *ability* to fix nitrogen.)
Capacity	The full amount that something can hold, contain, or receive. (The *capacity* of the beaker was 500 ml.)

ACCEPT, EXCEPT

Accept	*Accept* means to answer affirmatively, to receive something offered with gladness, or to regard something as right or true. (The scientists *accepted* his new theory.)
Except	*Except* is generally construed as a preposition meaning with the exclusion of or other than. It can also be a verb meaning to leave out. (Of the bacteria described, all are gram negative *except Streptococci.*)

ADMINISTER, ADMINISTRATE

Administer	(The drug was *administered* orally.) *Administration* is the noun of *administer*, which may lead to this common confusion between *administrate* and *administer*.
Administrate	*Administrate* means to manage or organize.

ACCURATE, PRECISE, REPRODUCIBLE

Accurate	*Accurate* means errorless or exact. It is often used in the sense of providing a correct reading or measurement. (The readings obtained with the spectrophotometer were *accurate*.)
Precise	*Precise* means to conform strictly to rule or proper form. (The value 5.26 is more *precise* than the value 5.3.)
Reproducible	*Reproducible* means something can be copied or repeated with the same results. (The experiment described by William et al. was *reproducible*.)

ADAPT, ADOPT

Adapt	*Adapt* is a verb and means adjust and make suitable to. (Most organisms *adapt* easily to minor changes in the environment.)
Adopt	*Adopt* is also a verb and means to accept or make one's own. (We *adopted* a new protocol for DNA isolation.)

ADVICE, ADVISE

Advice	*Advice* is a noun meaning suggestion or recommendation. (To measure dP/dt, we followed the *advice* of J. R. Boyd.)
Advise	*Advise* is a verb and means to suggest or to give advice. (Dr. Boyd *advised* us to follow his protocol.)

AFFECT, EFFECT

Affect	*Affect* is a verb and means to act on or influence. (The addition of Kl-3 to MZ1 cells *affected* their growth rate (i.e., it could have increased or decreased or induced)) *Affect* can also be a noun with a specialized meaning in medicine and psychology: an emotion. (People can experience a positive or negative *affect* as a result of their thoughts.)
Effect	As a noun, it means a result or resultant condition. (We examined the *effect* of Kl-3 on MZ1 cells.) As a verb it means to cause or bring about. (The addition of Kl-3 to MZ1 cells *effected* their growth rate (i.e., it had caused or brought about). She was able to *effect* a change in his attitude.)

AGGRAVATE, IRRITATE

Aggravate	When an existing condition is made worse, it is *aggravated*.
Irritate	When tissue is caused to be inflamed or sore, it is *irritated*.

ALLUDE, ELUDE, REFER, REFERENCE

Allude	*Allude* means to mention indirectly. (The authors of "Basic Medical Microbiology" only *allude* to brain abscesses in Chapter 10.)

Elude   *Elude* means to escape from or to escape the understanding or grasp of something. (The cause of the brain abscesses *eluded* us.)

Refer   *Refer* means to pertain or to direct to a source. (*Refer* to Barett et al. for more detailed listings. Questions *referring* to brain abscesses should be directed to a specialist in the field.)

Reference   *Reference* can be used as a noun, meaning a note in a publication referring the reader to another source (Do not forget to include *references* in the paper.) The term can also be used as a verb, meaning refer to (They *referenced* his work.)

ALTERNATELY, ALTERNATIVELY

Alternately   *Alternate* means every other one in a series. (Students *alternately* attended lectures on molecular biology and biochemistry every Saturday.)

Alternatively A choice between two or more mutually exclusive possibilities. (Students can major in biology or, *alternatively*, in chemistry.)

AMONG, BETWEEN

Among   *Among* means in a group of or the entire number of. It is used to express the relation of one thing to a group. (We discovered one black sheep *among* the white ones.)

Between   Use *between* with two items or more than two items that are considered as distinct individuals. (We found no marked differences *between* our results and those reported in Nature.)

AMOUNT, CONCENTRATION, CONTENT, LEVEL, NUMBER

Amount   Quantity that can be measured but not counted. (The total *amount* of yeast extract required for the medium was 12 g.)

Concentration The density of a solution or the amount of a specified substance in the unit amount of another substance. (The *concentration* of protein in the blood was 2.6 mg/ml.)

Content   A portion of a specified substance. (Soybeans have a high protein *content*.)

Level   Relative position or rank on a scale, often used as a general term for amount, concentration, or content. (Heart rate was at normal *level*. Protein *levels* (that is concentration) remained stable.)

Number   A quantity that can be counted. (The *number* of proteins in the aggregates varied.)

ANYBODY, ANY BODY, ANYMORE, ANY MORE, ANYONE, ANY ONE

Anybody   Refers to an unspecified person. Often used in place of everyone. (*Anybody* can get sick.)

Any body	A noun phrase referring to an arbitrary corpse or human form. (We found the head, but we did not see *any body*.) *This rule also applies to everybody, nobody, and somebody.*
Anymore	An adverb denoting time. (Doctors in the Western world don't prescribe chloramphenicol *anymore*.)
Any more	Used with a noun or as an indefinite pronoun. (We don't need *any more* measurements.)
Anyone	Refers to any person. Used like anybody and everyone. (*Anyone* can get sick.)
Any one	The two word form is used to mean whatever one person or thing of a group. (I would like *any one* of these apples.)

ASSAY, ESSAY

Assay	A test to discover the quality of something. (In this *assay,* a Pt electrode was used.)
Essay	A piece of writing, not poetry or a short story. (In this *essay*, she discussed her work in physical chemistry.)

AS, LIKE

As	*As* is a conjunction and is used before phrases and clauses. Rather than "*like* we just mentioned," say "*as* we just mentioned."
Like	*Like* is a preposition with the meaning of "in the same way as." (He was *like* a son to me.) Note the differences in the uses of *like* and *as*: *Let me speak to you as a father* (= I am your father and I am speaking to you in that character). *Let me speak to you like a father* (= I am not your father but I am speaking to you as your father might).

ASSUME, PRESUME

Assume	*Assume* means to take for granted or to suppose. It is usually associated with a hypothesis in scientific writing. (*Assuming* this hypothesis is correct, food should be supplemented with folic acid and vitamin B.)
Presume	*Presume* means to believe without justification. (Scientists *presume* a common ancestor for the two species.)

ASSURE, ENSURE, INSURE

Assure	*Assure* is to state positively and to give confidence and is used with reference to a person. (Let me *assure* you that we did not forget to add any enzyme.)
Ensure	*Ensure* means to make certain. (When you ligate DNA, you have to *ensure* that you do not forget to add the enzyme.)
Insure	*Insure* also means to make certain and is interchangeable with *ensure*. In American English, *insure* is widely used in

the commercial sense of "to guarantee financially against risk." (We insured our car.)

BECAUSE, SINCE

See SINCE, BECAUSE

BUT, AND, BECAUSE

But, and,
and because

Can be used to start a sentence as long as the sentence is complete. (Some bacteria can form endospores. *But E. coli and E. sakazakii* do not.)

CAN, MAY

Can

Means to be able to, to have the ability or capacity to do something. (Tetracyclin *can* be used to treat urinary tract infections.)

May

Indicates a certain measure of likelihood or possibility to do something. It also refers to permission. (Tetracycline *may* act by binding to a certain site of the bacterial ribosome.)

CAN'T, DIDN'T, HAVEN'T

These contractions cannot be used in scientific writing at all. Write them out instead: *cannot, did not, have not.*

COMPLIMENT, COMPLEMENT

Compliment

Compliment means praise. (John *complimented* Jean on her new dress.)

Complement

To complement means to mutually complete each other. (The protein bindin and its receptor *complement* each other.)

COMPRISE, COMPOSE, CONSTITUTE

Comprise

The conservative definition of comprise is to include or to contain. (The United States *comprises* many different states.) Avoid the phrase *is comprised of.*

Compose

Compose means to make up or to create something. Frequently used in the passive voice. (Water *is composed of* hydrogen and oxygen.)

Constitute

Constitute means to make a whole out of its parts, to equal or amount to. (Many different organisms *constitute* a habitat.)

CONSERVATIVE, CONSERVED

Conservative

Conservative implies a medical treatment that is limited or treatment with well-established procedures. (Due to complications, we selected *conservative* treatment for this case.)

Conserved	*Conserved* means to keep constant through physical or evolutionary changes. (DNA sequences may be *conserved* between species.)

CONTINUAL, CONTINUOUS

Continual	Repeatedly, occurring at repeated intervals, possibly with interruptions. (Measurements were hampered by *continual* interruptions.)
Continuous	Without interruption, unbroken continuity. (Our experiments were *continuously* interrupted means the experiments were started and then interrupted, and this interruption never stopped. The light spectrum is *continuous.*)

CONTRARY TO, ON THE CONTRARY, ON THE OTHER HAND, IN CONTRAST

Contrary to	Preposition meaning in opposition to. (*Contrary to* our expectations, addition of Mg^{++} did not alter our results.)
On the contrary	*On the contrary* is used when one says a statement is not true. It is a subjective statement that indicates opposition and is usually used only in spoken English. ("It's exciting!" "*On the contrary*, it's boring!")
On the other hand	Use *on the other hand* when adding a new and different fact to a statement. Rarely used in scientific English. (It's cold, but *on the other hand*, it's not raining.)
In contrast	*In contrast* is also used for two different facts that are both true, but it points out the surprising difference between them. It is an objective statement for a marked difference and can be used in scientific writing. (pH values increased for procaryotes. *In contrast*, no pH difference was observed in eucaryotes.)

DATUM, DATA

Datum	*Datum* is singular and means one result.
Data	*Data* is the plural form of *datum*. It should be used with the plural form of the verb. (Our *data* indicate that many experiments have to be repeated.) *The same applies for criterion, criteria, and medium, media.*

DESCRIBE, REPORT

Describe	Patients, persons, or cases are *described*. (We *describe* a patient with gynecomastia induced by omeprazole.)
Report	Cases and diseases are *reported*. (We *report* a case of omeprazole-induced gynecomastia.)

DIFFERENT FROM, DIFFERENT THAN

Different from	Correct expression. (The binding site of amoxicillin is *different from* that of penicillin.)

Different than	Incorrect.

DIE OF, DIE FROM

die of and die from	Persons and animals *die of,* not *from,* specific diseases.

DOSE, DOSAGE

Dose	A *dose* is a specific amount administered at one time or the total quantity administered. (Patients received an initial *dose* of 10 mg.)
Dosage	*Dosage,* the regulated administration of doses, is usually expressed as a quantity per unit of time. (Give a *dosage* of 0. 5 mg every 2 hours.)

ENHANCE, INCREASE, AUGMENT

Increase	A general term that means to become or to make greater. (Although we *increased* the amount of nutrients, the number of bacteria decreased.)
Augment	Means to make something already developed greater. (Addition of A greatly *augmented* the effect of B on C.)
Enhance	Means to make greater in value or effectiveness. In scientific writing, terms such as *increase* or *decrease* are preferred over *enhance* because they are much more precise.

ETC/SO ON, SO FORTH, AND THE LIKE

Etc.	Can only be used when the contents of a noninclusive list are obvious to the reader. It is an imprecise expression and should generally be avoided in scientific writing. Instead, use *such as* or *including* at the start of the list, and put nothing at the end of the list.
	Also avoid *and so on, and so forth,* and *and the like.*

EXAMINE, EVALUATE

Examine	Patients are *examined.*
Evaluate	Conditions and diseases are *evaluated.*

FARTHER, FURTHER

Farther	Refers to a physical distance. (You need to drive much *farther* than you think.)
Further	Refers to quantity or time. (The binding of X to Z needs to be examined *further.*)

FEWER, LESS

Fewer	Use *fewer* for items that can be counted (*fewer* cells, *fewer* patients).

Less	Use *less* for items that cannot be counted (*less* medication, *less* water). Exceptions include time and money (*less* than two years ago; Jack has less money than John).

FOLLOW, OBSERVE

Follow	A case is *followed*.
Observe	A patient is *observed*.
	To "follow up" on either approaches jargon, as does the term "follow-up study." However, in scientific writing, the use of the term *follow-up* is increasingly common, as is "follow-up study."

IMPLY, INFER

Imply	To *imply* is to suggest, indicate, or express indirectly. (These results *imply* that there is a key–lock mechanism between the enzyme and its receptor.)
Infer	To *infer* is to conclude or to deduce. (Looking at the interaction between the enzyme and its receptor, we can *infer* that a key–lock mechanism is used.)

INCIDENCE, PREVALENCE

Incidence	means occurrence or frequency of occurrence. (The *incidence* of macular degeneration is high in the elderly.)
Prevalence	means total number of cases of a disease in a given population at a specific time. (The *prevalence* of macular degeneration in the Western world in the year 2000 was 1 in 1000 individuals.)

INCLUDE, CONSIST OF

Include	Means partial; to be made up of, at least in part; to contain. *Include* often implies partial listing. (Various antibiotics, *including* streptomycin and tetracycline, bind to the bacterial ribosome.)
Consist of	Means to be made up of, to be composed of. Usually used when a full list is provided. (Macrolites *consist of*....)

INFECT, INFEST

Infect	Endoparasites *infect* or produce an infection.
Infest	Ectoparasites *infest* or produce an infestation.

INTERVAL, PERIOD

Interval	The amount of time between two specified instants, events, or states. (Optical density was measured at 10-min *intervals*.)
Period	An interval of time characterized by certain conditions and events. (Pterodactylus lived during the Jurassic *period*.)

IRREGARDLESS, REGARDLESS

Irregardless Does not exist in English. The correct form is *regardless*.

IT'S, ITS

It's Is a contraction of *it is* or *it has*. Preferably spell it out in scientific writing.

Its Means belonging to. (Bindin is an adhesive protein found at the tip of the sperm cell. *Its* structure is unknown.)

LATER, LATTER, FORMER, LAST, LATEST

Later Refers to time. (The fertilization membrane was not observed until *later* in the assay.)

Latter Means the second of two items. (*Hpa*II and *Msp*I both cut the recognition sequence CCGG, the *latter* also when the sequence is methylated.)

Former Means the first of two items mentioned. (*Hpa*II and *Msp*I both cut the recognition sequence CCGG, the *former* only when the sequence is not methylated.)

Last *Last* refers to the last item of a series. (The samples were pooled, incubated at room temperature for 10 min, and centrifuged at $300 \times g$ for 20 min. For the *last* step, 10 ml 80% ethanol was added.)

Latest *Latest* refers to the most recent item in a chronological series. (The *latest* book on molecular biology was published about one month ago.)

LAY, LIE

Lay To place or put something. (*Lay* is a transitive verb; that is, it needs an object.) Past tense is *laid,* past participle is *laid.* (Birds *lay* eggs.)

Lie To rest on a surface. (Lie is an intransitive verb, that is, it cannot take a direct object.) Past tense is *lay,* past participle is *lain.* (The tree line end where the lake *lies.*)

LOCATE, LOCALIZE

Locate To determine or specify the position of something. (Next, we *located* the vagus nerve.)

Localize To confine or to restrict to a particular place. (YT-31 was *localized* in the mitochondria.)

MEDIUM, MEDIA

Medium Singular noun. Needs a singular verb. (LB *medium* was prepared at room temperature.)

Media Plural form of medium. Needs a plural verb. (For the refined analysis, we used six different types of *media*.)

MILLIMOLE, MILLIMOLAR, MILLIMOLAL

Millimole   (mmol); an amount, not a concentration. (A 1 *millimolar* solution contains 1 *millimole* of a solute in 1 liter of solution (or, a 1 *mM* solution contains 1 *mmol/liter* of solution.)

Millimolar (mM); a concentration, not an amount. (A 0.5 *millimolal* solution contains 0.5 *mmol* of a solute in 100 g of solvent. The final volume may be more or less than 1 liter.)

Millimolal A concentration, not an amount.

MUCUS, MUCOUS

Mucus   The noun. (*Mucus* is a viscous substance secreted by the *mucous* membranes.)

Mucous The adjective meaning containing, producing, or secreting mucus.

MUTANT, MUTATION

Mutation   A *mutation* is an alteration in the primary sequence of DNA. (A *mutation* can be mapped, but a *mutant* cannot.)

Mutant Refers to a strain of organism, population, allele, or gene that carry one or more mutations. (A *mutant* has no genetic locus, only a phenotype.)

NECESSITATE, REQUIRE

Necessitate   *Necessitate* means to make necessary or unavoidable. (The treatment may *necessitate* certain procedures.)

Require *Require* means to have a need for or to demand. (A patient *requires* treatment.)

NEGATIVE, NORMAL

Negative   Tests for microorganisms and reactions may be *negative* or positive.

Normal Observations, results, or findings are *normal* or *abnormal*. (The patient's behavior was *normal*.)

OPTIMAL, OPTIMUM

Optimal Adjective meaning most favorable; never used as a noun. (The *optimal* (or *optimum*) temperature for X was 28°C.)

Optimum Noun meaning the most favorable condition for growth and reproduction; often used as an adjective. (X reached its *optimum* 30 hours after induction.)

OVER, MORE THAN

Over   *Over* can be ambiguous. (The cases were followed up *over* two years). Instead, use "more than." (The cases were followed for *more than* two years.)

More than   Preferred.

PERCENT, PERCENTAGE

Percent   Means per hundred. Written as % together with number in scientific writing. (The yield of the gene product was less than 5%.)

Percentage   Is a general part or a portion of the whole. Usually with a singular verb. (A small *percentage* of the crop was spoiled.)

PARAMETER, VARIABLE, CONSTANT

Parameter   Means a constant (of an equation) that varies in other settings of the same general form. Parameters are a set of measurable factors, such as temperature, that define a system and determine its behavior. (Changing the *parameters* of the system will result in a different outcome.)

Variable   A *variable* is a quantity that can change in a given system. (In the equation $y = ax + b$, x is the *variable*.)

Constant   A *constant* is a quantity that is fixed. (π is a *constant*.)

PERTINENT, RELEVANT

Pertinent   Means having logical, precise relevance to the matter at hand. (These assignments are *pertinent* to understand the class material.)

Relevant   What relates to the matter or subject at hand. (He performed experiments *relevant* to is research.)

PRINCIPAL, PRINCIPLE

Principal   *Principal* can be either a noun meaning a person or thing of importance or an adjective meaning main or dominant. (The principal reason for not observing any gas formation was low temperature.)

Principle   *Principle* can only be a noun and means law or general truth. (This manual presents many writing *principles*.)

QUOTATION, QUOTE

Quotation   *Quotation* is a noun meaning a passage quoted. (Indicate a *quotation* with quotation marks.)

Quote   *Quote* is the verb and means to repeat or to cite. (Sentences copied from other sources should be *quoted*.)

QUANTIFY, QUANTITATE

Quantify   Often interchanged and a matter of personal preference. Both are used with the meaning to determine or measure the quantity of something.

Quantitate   *Quantitate* is preferred if one wants to emphasize that something was measured *precisely*.

RATIONAL, RATIONALE

Rational   An adjective meaning able to reason. (This is not a very *rational* thing to do.)

Rationale   A noun meaning basis or fundamental reason. (The *rationale* behind our theory is that cancer incidences are ever increasing.)

REGIME, REGIMEN

Regime   A *regime* is a form of government.

Regimen   *Regimen* means a regulated system intended to achieve a beneficial effect. When a system of therapy is meant, *regimen* is the correct term.

REMARKABLE, MARKED

Remarkable   Is often incorrectly used to indicate a change that is notable but not significant. The correct word is *marked*. (There was a *marked* increase in binding.)

Marked   See *remarkable*.

REPRESENT, BE

Represent   Means to stand for or to symbolize. (Each data point *represents* the average of five measurements.)

Be   Equal, constitute. (Mercury *is* a toxic substance.)

SINCE, BECAUSE

Since   Use this word only in its temporal sense and not as a substitute for *because*. (We have had nothing but trouble *since* we moved here.)

Because   If you want to indicate causality, use *because*. (The reaction rate dropped *because* temperature dropped.)

SYMPTOMS, SIGNS

Symptoms   *Symptoms* apply to people. (The patient displayed no *symptoms* of the disease.)

Signs   *Signs* apply to animals. (A *sign* that the dog was sick was that she did not eat anymore.)

THAN, THEN

Than   *Than* is a conjugation that introduces an unequal comparison. (Birds are closer related to dinosaurs *than* to mammals.)

Then   Means next in time, space, or order. (The samples were centrifuged at 100 *x g*. *Then* the pellet was resuspended.)

TOXICITY, TOXIC

Toxicity   *Toxicity* is the quality, state, or degree of being poisonous.

Toxic   *Toxic* means poisonous. (Mercury is *toxic* for most organisms.)

VARYING, VARIOUS

Varying	*Varying* means changing. (Precise measurements could not be taken because of *varying* levels of humidity.)
Various	*Various* means different. (The crystals were of *various* sizes.)

VIA, USING

Via	*Via* means by way of. (The students went to the lab *via* the ice cream parlor.)
Using	*Using* means to put into service and by means of. (We isolated the plasmid DNA *using* a commercially available kit.)

WHICH, THAT

	Sometimes the words can be used interchangeably. More often, they cannot.
Which	Use *which* with commas for nondefining (nonessential) phrases or clauses. (Dogs, *which* were treated, recovered.)
That	Use *that* without commas for essential phrase or clause. A phrase or clause introduced by *that* cannot be omitted without changing the meaning of the sentence. Such essential material should not be set off with commas. (Dogs *that* were treated with antidote recovered.)

WHILE, WHEREAS, ALTHOUGH

While	*While* indicates time and temporal relationship. It means at the same time that. It is often incorrectly used instead of *although* or *whereas*. (Experiments were performed *while* patients were sleeping.)
Whereas	*Whereas,* often the word the writer intended, means "when in fact" and "in view of the fact that" (*Whereas* X increased, Y decreased.)
Although	*Although* is a conjunction meaning despite the fact that or even though. (*Although* the association between breast cancer risk and variant alleles varied slightly, the *p* values were not statistically significant.)

USE, UTILIZE

Use	Generally, *use* is the preferred term.
Utilize	*Utilize* means to find a new, profitable, or practical use for something. Viewed as an unnecessary and pretentious substitute for use.

Appendix B

MS WORD BASICS AND TOP 20 MS WORD TIPS

The following instructions apply to MS Word 2010 for Windows (PC version). On the Mac, most commands are the same but may have to be accessed differently.

MS WORD BASICS

In the screen display of the command area for Word 2010, the **Ribbon** displays the commands in task-oriented **groups** in a series of **tabs** (Figure B.1). Additional commands within the groups can be accessed by clicking on the **Dialog Box Launcher**. Clicking on the **File Tab/ Backstage Button** in MS Word displays a drop down menu containing a number of options, such as *open, save, send, recent,* and *print.* You can customize the Quick Access Toolbar to contain buttons for tasks you perform frequently, such as save, print preview, and undo move. The window size and screen resolution affect what you see in the command area of the program.

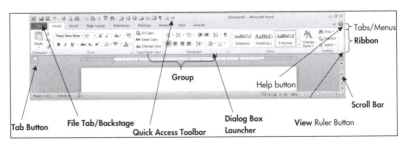

Figure B.1 Command area in MS Word

TOP 20 MS WORD TIPS

The following instructions apply to MS Word 2010 for Windows (PC version). On the Mac, most commands are the same but may have to be accessed differently.

Tip 1.   If you do not know how to perform a specific function in Word, click on the **help button**, which is a white question mark in a blue round field on the top right corner of the command area, and type in the key word or question you have. The help function provides you with different options to narrow down your specific concerns and give you a step-by-step explanation to follow.

Tip 2.   If you activate the **Show/Hide button** (⁋), which looks like a big π or pound sign in the Home tab Paragraph group, you can see the formatting symbols in your document as you work with it. This will enable you to identify spacing issues, including inconsistent spacing between words, hard page breaks versus sections breaks, and so on.

Tip 3.   To apply the same formatting of an already formatted text section to another section, use the **Format Painter**, which can be found in the Home-Clipboard group. Highlight the source text, click the format painter button and then select the text to which you want to apply the formatting.

Tip 4.   Use the **Print Preview** screen to see what your document looks like before you print it. To select Print Preview, click on the File Tab/Backstage button on the top left corner of the screen, then select Print and Print Preview automatically is displayed on the MS Office Backstage view.

Tip 5.   To write something as **super- or subscript**, highlight the letter or symbol you would like to convert to super or subscript. Click on the applicable button in the Home-Font group (x_2 for subscript, x^2 for superscript).

Tip 6.   If you need to add a special character, mathematical symbol, and/or foreign letter (such as α, β, or μ), choose the Insert tab and then the **Symbol** group. Select *more symbols* and then the symbol/character you want followed by *insert*. If you go back to that same dialog box, Word displays the symbols you have most recently used.

Tip 7.   To adjust **line spacing within paragraphs**, highlight the text for which you want to adjust the line spacing. Then, in the Home-Paragraph group, select the line spacing button and choose the line spacing option (1.0 for single, 2.0 for double, a specific value in between, etc.) that you would like.

Tip 8.   To adjust **line spacing before and after paragraphs**, click anywhere within or highlight the paragraph and go to the Home-Paragraph group. Click on the lower right arrow and go to the Spacing section in the dialog box. Click on the Before and/or After arrows until you get to 0 pt for both, and then click OK.

Tip 9.   To turn the **ruler on and off**, click on the View Ruler button on the right side just above the scroll bar.

Tip 10. To **bullet or number text**, click within the paragraph that you would like to bullet or number and then select the bullet or numbering you want from the Bullets or Numbering button in the Home-Paragraph group. If the numbered list does not start where you want, click the small arrow next to the numbering button, select *set numbering value* and choose the value you like.

Tip 11. Just below the main Ribbon on the left side is **Tab button**. This button designates the type of tab: left, center, right, decimal, etc. By clicking on the button, you can scroll through the different options. To set tabs for a particular section of text, first click in the line or paragraph you want to format. Then, scroll through the tab choices until you find the tab you want. Click on the horizontal ruler above the document at the point where you want to insert the tab, and the symbol for that tab will appear. You can click on the tab symbol on the ruler and slide it left or right as needed. You can use a combination of tabs to left-justify some text and right-justify the time at the end of the line. Using the right-justify tab is the only way to ensure that text on the right margin will be properly aligned there. (If you want to get rid of a tab, simply go to the text where the tab has been applied, click on the tab symbol on the ruler, and drag it off the ruler completely.)

Tip 12. To make a given **text section narrower or wider** than the normal page width in the rest of the document, highlight the applicable text and drag the left and/or right indents on the horizontal ruler above the page.

Tip 13. To **insert page numbers** that are automatically sequentially numbered, go to the Insert tab, chose Page Number within the Header & Footer group. The drop-down menu will allow you to select the placement and format of the page number. You can also designate which number to start with by selecting Format Page Number. If the formatting of the Word-inserted page numbers is not to your liking, you can simply change it as you would any text after double clicking it.

Tip 14. To **insert a table** in Word, go to Insert-Table and either highlight the rows and columns you need in the grid, or go to the Insert Table choice further down in the drop-down menu, type the number of rows and columns you want, and click OK. Note: There are advantages and disadvantages to using Word or Excel for tables. Word tables will give you more formatting options, but Excel tends to be easier to work with numbers, particularly where calculations are involved. (Note: You can do a simple sum in a Word table [in the Table Tools–Layout–Data group], using the Formula button.)

Tip 15. You copy a spreadsheet, chart, or picture from another Microsoft application (Excel, Visio, etc.) and paste it (under **Paste Special**) as an object into MS Word. You will be able to modify the object in your Word document, without going back to the original program.

Tip 16. To **crop a picture or object** in word, left-click the picture, which brings up the Picture-Format group. Click on Format, and then on Crop in the Size group. This will create a broken border around the picture. Position the cursor just tangent to one of the border segments, and then the curser will change to a "handle" you can use to adjust the size of the picture as you see fit. Once you move the cursor away from the picture, the picture will remain the cropped size, and you can then enlarge it if necessary. (Note that the original, full picture/object is still available from the cropped version if you want to undo this.)

Tip 17. CNTRL+Enter allows you to insert a **page break** at the place the cursor is located.

Tip 18. To **wrap text around pictures and objects**, click on the picture or object, and a Picture Tools toolbar will appear. In the Picture Tools–Format group, click on Text Wrapping and select Tight or Square.

Tip 19. **Tracked changes and comments** allow you to see edits and/or comments that others make to the document. You can then accept or reject each of the changes and respond to the comments as needed. To turn on tracked changes, go to the Review tab, and click on the Track Changes button in the Tracking group.

Tip 20. To **insert comments**, go to the Review tab, and click on the Insert Comment button in the Tracking group.

MS WORD SPECIAL CHEAT SHEET

Keyboard Shortcuts

Three hyphens (---) + ENTER	Normal line across the page
Three underscore (___) + ENTER	Bold line across the page
Three equal signs (===) + ENTER	Double line across the page
Three hashes (###) + ENTER	Thick line with thin lines above and below across the page

Appendix C

EXCEL BASICS AND TOP 20 EXCEL TIPS

The following instructions apply to MS Excel 2010 for Windows (PC version). On the Mac, most commands are the same but may have to be accessed differently.

EXCEL BASICS

In the screen display of the command area for Excel 2010, the **Ribbon** displays the commands in task-oriented **groups** in a series of **tabs** (Figure C.1). Additional commands within the groups can be accessed by clicking on the **Dialog Box Launcher**. Clicking on the **File Tab/Backstage Button** in Excel 2010 is similar to clicking on that in MSWord or PowerPoint. It displays a drop down menu containing a number of options, such as *open, save, send, recent,* and *print*. You can customize the Quick Access Toolbar to contain buttons for tasks you perform frequently, such as save, print preview, and undo move. The window size and screen resolution affect what you see in the command area of the program.

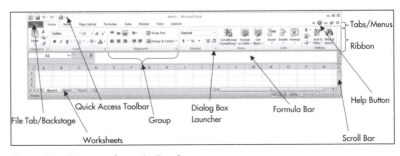

Figure C.1 Command area in Excel

TOP 20 EXCEL TIPS

Tip 1.   If you do not know how to perform specific functions in Excel, click on the **help button,** which is a white question mark in a blue round field on the top right corner of the command area, and type in the key word or question you have. The help function of Excel will provide you with different options to narrow down your specific concerns and give you a step-by-step explanation to follow.

Tip 2.   You can use the arrow keys *(left, up, down, right)* or the PgUp and PgDn keys to **navigate to a specific cell** instead of the mouse. Enter will go down one cell, and the Tab key will go to the next cell to the right. Ctrl + Home key will go to the first cell. Ctrl + End key will go to the last cell.

Tip 3.   To quickly copy formatting from one part of a sheet to another, or to another sheet in the same workbook, use the **Format Painter,** which can be found in the Clipboard group of the Home menu. To use Format Painter to copy formatting options to another cell(s), add all the formatting options you want to use to at least one cell, select the cell, and then click on the Format Painter icon on the formatting toolbar. After that, click on the cell that you want to copy the formatting to.

Tip 4.   Excel can **follow a sequence** you typed in. For example, if you typed in the word January, you can use a technique to have the next cell display February, then March and so forth. To create such a series, click on the cell where you typed in the first item in the series (January). Notice on the bottom right of the cell that there is a little square, called a handle. Put the mouse cursor over the little square. The cursor will turn into a + sign. Click the left mouse button and while continuing to hold down the button move your mouse down *(or to the right)* to the cell to which you would like to extend the series. Let go of the mouse button and the series will be displayed. This technique works for time, numbers, months, calendar, quarters, and a few other series. If you double click the handle, Excel will fill in an entire series for a table column automatically.

Tip 5.   All **formulas and functions** begin with the equal sign. For example, to add x and y together, the formula would be =x+y, where x and y can be numbers or cells (A1, B1, etc.). See below for specific formulas to use for diverse mathematical functions.

Tip 6.   **Ctrl+Roll Mouse wheel** zooms in and out within the Excel display window.

Tip 7.   If you want to **add up items in a column and/or row,** highlight the items you want to add, plus one empty cell. For example, if you had numbers in cells A1 through A5, highlight A1 through A6. On the Home tab in the Editing group, select the Greek Symbol Σ. Clicking on that symbol will display the sum of the numbers in the empty cell that you highlighted.

Tip 8. If you had a **number that you wanted to display** as a percentage, decimal, or dollar amount, highlight the cell and on the Home tab, select the percent (%), decimal (.), or dollar ($) sign in the Number group. Clicking on the corresponding symbol will display the number in the cell as a percent, decimal, or dollar amount.

Tip 9. For a **shortcut of the most common features** to use in Excel, highlight the cell(s) and click on the right mouse button. A window will open and the most commonly used features will be displayed in the menu.

Tip 10. To **insert or delete a row or column**, click on the letter for inserting a column, or number for inserting a row. After clicking on the number or letter, click the right mouse button and select Insert from the menu option.

Tip 11. If you like to **use Windows Calculator** to perform quick calculations while in Excel, you can save time by adding it to your Quick Access Toolbar. To add a calculator, click Customize Quick Access Toolbar to the right of your Quick Access Toolbar, and then choose "More Commands". On the left-hand side choose "Commands Not in the Ribbon" from the drop down menu, and you'll see Calculator in the list on the left pane. Just click the Add button to add it to the toolbar.

Tip 12. To **place Excel data into MS Word**, you can

a. Copy and paste the table from Excel into the word file. This method is good for one time presentation. The main limitation of this method is that you would not be able to modify the content of the table later if you wish to.

b. Embed Excel data into your Word document. Embedding the Excel data directly in Word will allow you to modify the data directly in Word instead of making a sheet all over again in Excel when you want to modify some numbers. However, the embedded tablet will not be linked to the original Excel file anymore; that is, if you modify your source file, then your Excel sheet in the Word document would not get updated automatically. To embed Excel data into a Word document,

- Select and copy the range of data in Excel that you want to embed in your Word document
- Go to the Word document and place your cursor at the position where you want to put the Excel data. Then, in Clipboard group of the Home tab choose Paste Special from the special dialogue box below the Paste command.
- Select Microsoft Office Excel Worksheet Object from the list of options.

Once it is embedded, you can then click on a table and will be able to edit the contents because it will open in Excel. The main limitation of this method is that it makes the Word document a very large file.

c. **Link the Excel data to your document**. If you link your Excel worksheet in Word to the Excel source file, the worksheet in Word gets updated automatically whenever you modify your original Excel sheet. To link your Excel data, follow the exactly same steps under b. However, instead of choosing the paste option, choose "Paste Link" from the Paste Special dialogue box.

d. If you want to **insert an empty worksheet,** you can choose the Insert tab and then the Table group and there click the small arrow to choose Excel spreadsheet from the menu.

Tip 13. To **hide columns and rows** and present a sheet that focuses on just the work area, select the row or column header from the row or columns to be hidden. Right click within the selected rows or columns and select *hide*. To unhide the column or row, select the row or column before and after the hidden ones, right click, and select *unhide*.

Tip 14. To **show or hide formulas** when you are working in an Excel worksheet, you can alternate between viewing the values in the cells and displaying the formulas by pressing CTRL+` (single left quotation mark). Note: If you're having trouble finding the single left quotation mark, it's on the same key as the "~" symbol.

Tip 15. **To select a large data set**, select the top left-hand cell and then use CTRL+SHFT+↑/↓ to move to top/bottom cell in a column and row.

Tip 16. Use Ctrl+↑/↓ to jump to top/bottom cell in a column.

Tip 17. When reading or inputting data into a large table, it is useful to **freeze the horizontal or vertical panes**, or both, to orient yourself and keep columns and rows visible while the rest of the worksheet scrolls up and down. To freeze horizontal panes, go to the row below it, select the View tab, and in the Windows group, click *freeze panes*. The row header will be "frozen" when you scroll downwards. To freeze vertical panes, go to the column on the right and repeat the same procedure. To freeze both vertical and horizontal panes, select the pane below and to the right, and repeat the procedure. To unfreeze panes, select *unfreeze* panes in the Windows group of the View tab.

Tip 18. To **turn on the gridlines** so that they will be printed, select the worksheet or worksheets that you want to print. On the Page Layout tab, in the Sheet Options group, select the Print check box under Gridlines.

Tip 19. You can **sort your data columns and rows** in various ways, from alphabetical to values and colors. To ensure that corresponding rows and columns stay together as such, be sure to click expand the selection after choosing custom sort from the sort and filter command in the Editing group of the home menu.

Tip 20. To **create a Chart** with the push of a keyboard button, select the data you want to include in the chart, including the labels. Press F11 key or the Alt+F1 key.

EXCEL SPECIAL CHEAT SHEETS

Keyboard Shortcuts

Ctrl+'(apostrophe)	Copies contents of cell above
Ctrl+Pg up/Pgdown	Move to next work sheet
Ctrl+↑/↓	Move to top/bottom cell in a column
Ctrl+Shft+↑/↓	Select to top/bottom of column
Alt+Enter	Wrap text in a cell
Click, then SHFT+click	Selects cells between clicks
Ctrl+Click – while holding -over different cells	Selects all clicked items
CTRL+` (single left quotation mark)	Shows formula for the cell
CTRL+;	Displays today's date

Fixing Common Error Messages

Error messages start with a pound (#) sign.

#####	If you see rail road tracks, your column is too narrow. Solution—widen the column.
#REF!	Your formula refers to a cell that no longer exists, due to a change in the worksheet
#DIV/O!	You are dividing by an empty cell or zero. Solution—fix the formula's denominator.
#NUM!	A formula or function contains invalid numerical values.
#VALUE!	Your formula includes cells that contain different data types, such as text instead of a number.
#NAME?	Your formula contains text that Excel does not recognize, such as a type or missing punctuation.
CIRCULAR	When a formula refers to a cell in which it is located (e.g., A1+10 CANNOT BE USED in A1=10)

Formulas

Addition of two cells	=A2+B3
Addition of a constant	=A2+10
Addition of a row of cells	=SUM(A1:C1)
Addition of a column of cells	=SUM(A1:A3)
Addition of a range of cells	=SUM(B1:C3)
Addition of scattered cells	=SUM(A2,B1,C3)
Subtraction of a constant	=A1−10
Subtraction of a cell	=B2−B1
Multiplication by a constant	=A1*10
Multiplication of two cells	=A1*B3
Multiplication by a percent	=A1*.40 or =B2*40%
Division by a constant	=A1/10
Division by a cell	=A1/B2

Exponentiation (square)	=A1^2
Exponentiation (cube)	=A1^3
Square root	=SQRT(A1)
Cube root	=A1^(1/3)
Increasing by a percentage	=A1+(A1*.4) or =A1*1.4 or =A1+(A1*40%)
Decreasing by a percentage	=A1−(A1*.06) or =A1*.94 or =A1−(A1*6%)
Calculate a percentage (part/Sum)	=A1/D3
Average of a column	=AVG(A1:A3)
Average of a row	=AVG(A1:C3)
Average of a range	=AVG(A1:B3)

GRAPHING WITH EXCEL

Tip 1. To **create a customized chart** in Excel, select the entire data range (including the headings). Then, choose the Insert tab followed by the chart type from the Charts group. Choose the chart subtype.

Tip 2. **Changing the chart type** of your chart is easy. Right click the chart, and then click Chart Type.

Tip 3. You can **change the chart size** by dragging its corners. You can also change its location by dragging it to the desired place.

Tip 4. You can **format text and numbers on a chart** by right clicking them and using the mini toolbar

Tip 5. To **format areas in a chart**, select them with a click and then use the Shape Fill group from the Format tab.

Tip 6. To **reshape an axis**, right click on the axis and then choose format axis from the menu. Ensure that the axis options function is selected on the left pane. Choose Fixed on the minimum and maximum row, and type the axis starting and ending number. The select fixed on the Major unit row and enter the axis step between numbers. Click on Close.

Tip 7. To **plot a semilog graph** in Excel, double click the y axis, then from the Scale tab select Logarithmic scale.

Tip 8. To **create a log–log graph** in Microsoft Excel, you first need to create an XY (scatter) graph (see tip1 above). Select Logarithmic scale from the Scale tab.

Tip 9. If your chart data contains one or more empty cells, you will find that if you plot the data the empty cells will be skipped, leaving a break in your line. To **configure how empty cells and hidden data are plotted**, click the chart and choose the Chart Tools, Design tab. Click the Select Data button and select Hidden and Empty Cells button. Choose to show empty cells as either Gaps or Zero or select the Connect data points with line option. Click OK to apply the changes to your chart.

Tip 10.   To **add a trendline** or best-fit line to your graph, move the mouse cursor to *any data point* on the chart where you wish to create the trendline and press the left mouse button. All of the data points should be highlighted. While the mouse cursor is still on any one of the highlighted data points, press the right mouse button, and click on **Add Trendline** from the menu that appears. From within the "Trendline" window, click on the box with the type of fit you want (e.g., Linear). Click on Options at the top of the "Trendline" window. Click in the checkbox next to "Display Equation on Chart" and the checkbox next to "Display R-squared Value on Chart". Do not click on the checkbox next to "Set Intercept = 0". Click OK. A line, an equation, and an R-squared value should appear on the graph.

Appendix D

POWERPOINT BASICS AND TOP 20 POWERPOINT TIPS

POWERPOINT BASICS

In the screen display of the command area for PowerPoint 2010, the **Ribbon** displays the commands in task-oriented **groups** in a series of **tabs** (Figure D.1). Additional commands within the groups can be accessed by clicking on the **Dialog Box Launcher**. Clicking on the **File Tab/Backstage Button** in PowerPoint displays a drop down menu containing a number of options, such as *open, save, send, recent,* and *print.* You can customize the Quick Access Toolbar to contain buttons for tasks you perform frequently, such as save, print preview, and undo move. The window size and screen resolution affect what you see in the command area of the program.

TOP 20 POWERPOINT TIPS

The following instructions apply to PowerPoint for Windows (PC version). On the Mac, most commands are the same but may have to be accessed differently.

Tip 1. If you do not know how to perform a specific function in PowerPoint, click on the **help button**, which is a white question mark in a blue round field on the top right corner of the command area, and type is the key word or question you have. The help function provides you with different options to narrow down your specific concerns and give you a step-by-step explanation to follow.

Tip 2. To apply the same formatting of an already formatted text section to another section, use the **Format Painter**, which can be found in the Home-Clipboard group. Highlight the source text, click the format painter button, and then select the text to which you want to apply the formatting.

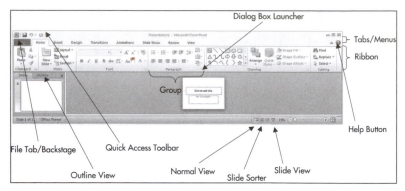

Figure D.1 Command area in PowerPoint

Tip 3. If you need to **add a special character**, mathematical symbol, and/ or foreign letter (such as α, β, or μ), choose the Insert tab and then the **Symbol group**. Select *more symbols* and then the symbol/character you want followed by *insert*. If you go back to that same dialog box, Word displays the symbols you have most recently used.

Tip 4. To adjust **line spacing within paragraphs**, highlight the text for which you want to adjust the line spacing. Then, in the Home-Paragraph group, select the line spacing button and choose the line spacing option (1.0 for single, 2.0 for double, a specific value in between, etc.) that you would like.

Tip 5. To adjust **line spacing before and after paragraphs**, click anywhere within or highlight the paragraph and go to the Home-Paragraph group. Click on the lower right arrow and go to the Spacing section in the dialog box. Click on the before and/or after arrows until you get to 0 pt for both, and then click OK.

Tip 6. To **bullet or number text**, click within the paragraph that you would like to bullet or number and then select the bullet or numbering you want from the Bullets or Numbering button in the Home-Paragraph group. If the numbered list does not start where you want, click the small arrow next to the numbering button, select *set numbering value*, and choose the value you like.

Tip 7. To **crop a picture or object in** PowerPoint, left click the picture, which brings up the Picture-Format group. Click on Format, and then click on Crop in the Size group. This will create a broken border around the picture. Position the cursor just tangent to one of the border segments, and then the curser will change to a "handle" you can use to adjust the size of the picture as you see fit. Once you move the cursor away from the picture, the picture will remain the cropped size, and you can then enlarge it if necessary. (Note that the original, full picture/object is still available from the cropped version if you want to undo this.)

Tip 8. To **turn on and off the ruler and gridlines** in order to help you to place text and objects in PowerPoint, right click in the empty space on a slide and select ruler and/or grid and guides from the dialog box. Click ok.

Tip 9. If you want to **align two or more objects or text boxes** relative to each other, go to the Drawing group in the Home tab and select Arrange. In the dialog box, choose Align and then select the alignment you want (align left, align top, etc.)

Tip 10. To **insert a shape**, including a ring, bracket, or arrow, select the shape of choice from the Drawing group in the Home tab or from the Illustrations group under the Insert tab. Then drag out the shape to the desired size. To modify the shape, click on the Format tab under drawing tools. Here you can change the shape fill color (or no fill), outline color, and shape effect. You can also adjust these under the Drawing group in the Home tab.

Tip 11. To **animate text and objects** in PowerPoint go to the Animations tab and select the animation in the Animations group. An Animation pane dialog box will appear on the left side of the screen. To animate an object, select the object, and then select trigger in the Advanced Animation and timing and order in the Timing group. Fine tune the animation by right clicking on the animation in the Animation pane and selecting Effect Options.

Tip 12. To **make text bullets appear one-by-one**, select the text in the Animation pane, then right click and select Effect Options Text Animation Select by first-level paragraph.

Tip 13. To **hyperlink a slide** to another, or an object within a slide to another slide or file, right click on the text or object, word or design element, then select Hyperlink from the dialog box. Choose Place in this document (or other address), then select the slide or file.

Tip 14. **Backgrounds** can be added to individual slides or all slides in the presentation. Backgrounds for slides can be solid colors, gradient colors, textures, or pictures.

    a. *Using the Design Tab on the Ribbon*

- Click on the *Design* tab of the ribbon which contains the features you will need to add or change a background in PowerPoint.
- Click on the Background button on the right end of the ribbon. This will open the *Format Background* dialog box.

    b. *Right Clicking on the Slide*

- Right click on a blank area of the slide.
- Choose Format Background . . . from the shortcut menu. This will open the *Format Background* dialog box.

Tip 15. To **add photos or documents** to your slides, you can copy and paste them from a file folder or from another program that has an image showing, or you can use the PowerPoint program menus to insert any number of different items. Pictures can then also be formatted using the Format tab of the Picture Toolbox.

Tip 16. To **resize an image without distorting** it, hold down the Shift key while grabbing the corner of the image. If the image is too large for the slide, you can drag it or scroll until you can see a corner, then resize.

Tip 17. To **change Views** in PowerPoint, you have several options (see also Figure D.1):

- *Normal View* is the main working window in the presentation. The slide is shown full size on the screen.
- *Slide Sorter View* is a window in PowerPoint that displays thumbnail versions of all your slides, arranged in horizontal rows. This view is useful to make global changes to several slides at one time. Rearranging or deleting slides is easy to do in Slide Sorter view.
- *Slide View* shows a full screen version of your slides as they are projected onto a screen.

Tip 18. You can **adjust the order of objects** in PowerPoint by selecting one of the commands in the Order Objects Menu within the Drawing-Arrange group in the Home tab.

Tip 19. To **print more than one slide per page**, select Print from the File Tab/Backstage Button, and then choose Handouts or Notes pages from the Print Settings options. Click Print.

Tip 20. To **paste text or an object or slide without losing its original layout and format**, paste the text or object into the slide, and then look for the Paste Options button that appears close to the inserted text, object, or slide. Left click this button and select Keep Source Formatting.

POWERPOINT SPECIAL CHEAT SHEET

Keyboard Shortcuts

ESCAPE	Ends the presentation mode
CTRL+B; lowercase b; or full stop (.) during presentation	Blackens the screen during a presentation; click it again to continue
CTRL+W	Whites the screen during a presentation; click it again to continue
H	Go to hidden slide
Holding SHIFT while dragging a shape	Allows you to drag the shape in only one direction to help in alignment
Holding SHIFT while drawing a shape	Makes shape perfectly symmetrical

Appendix E

MS OFFICE CHEAT SHEET

F1	Help
F4	Repeats the last action
F12	Save As
Ctrl+Home	Beginning of document
Ctrl+End	End of document
Ctrl+Roll Mouse Wheel	Zoom in and out
CTRL+C	Copy
CTRL+X	Cut
CTRL+F	Find
CTRL+V	Paste
CTRL+P	Print
CTRL+S	Save
Ctrl+Pg up/Pgdown	Move to next worksheet
Ctrl+↑/↓	Move to top/bottom cell in a column
Ctrl+Shft+↑/↓	Select to top/bottom of column
Ctrl+Click—while holding over components	Selects all clicked items

Answers to Problems

Chapter 2 Literature Sources

Problem 2-1

Ostracodes are small bivalved Crustacea that form an important component of deep-sea meiobenthic communities along with nematodes and copepods (10). Crustaceans are dense and diverse in the deep sea and are one of the most representative groups of whole deep-sea benthic community (**10, 11**). Pedersen et al. (**12**) as well as Jackson et al. (**13, 14**) reported that ostracode species have a variety of habitat and ecology preferences (e.g., infaunal, epifaunal, scavenging, and detrital feeders), representing a wide range of deep-sea soft sediment niches. Furthermore, Ostracoda is the only commonly fossilized metazoan group in deep-sea sediments (**reference**). Thus, fossil ostracodes are considered to be generally representative of the broader benthic community. The distribution and abundance of deep-sea ostracode taxa in the North Atlantic Ocean are influenced by several factors, among them, temperature, oxygen, sediment flux, and food supply (**14, 15**). Several paleoecological studies suggest that these factors influence deep-sea ecosystems over orbital and millennial timescales (**1, 16**).

Problem 2-2

Droughts can occur in almost all climates, although their onset, severity, and frequencies vary widely. Droughts may be little noticed, if at all, in arid areas but result in dramatic loss of crops and water shortage in others. Droughts result when it has not rained enough over an extended period of time, usually a season or more. High temperatures and wind can add to their onset and lengths. Because drought characteristics are much disputed among scientists and politicians, not much progress in drought management has been seen.

Problem 2-3

The deadly H5N1 avian influenza virus could spark a global pandemic that could kill tens of millions once it acquires the ability to spread in humans. Since 2003, H5N1 has been spreading in an ever-increasing number of countries, from Asia to Africa and Europe. While outbreaks have surfaced in new countries in 2007, they have been more or less continuous in others, such as in Indonesia and in Nigeria. Although the virus is passed on to humans mainly through domesticated birds, it is unclear what role wild birds play in its ever-widening circles. To shed more light on their role in the spread of the virus, lightweight transmitters could be used to track migratory birds by satellite.

Problem 2-4

Version A would be considered plagiarized. Here, sentences have simply been rearranged, and a few words have been substituted. Overall, Version A is too close to the original. In contrast, Version B is paraphrased. It not only differs very much from the original, it also quotes the original source.

Chapter 3 Fundamentals of Scientific Writing, Part I—Style

Problem 3-1

1. Plants were kept [**at 0°C? or –20°C**] overnight.
2. [**10%**, 12 %, 6] of the discovered exoplanets have an orbital period of less than 5 days.
   POINT: "Some" is imprecise. How much is some? Use a quantitative term.
3. The afterglow of the glowing blast wave was [50%, 120%] brighter than we expected.
   POINT: "Markedly" is imprecise. How much is markedly? Use a quantitative term.

Problem 3-2

1. support
2. used
3. are due to/are caused by different conditions.
4. To...
5. For example,...

Problem 3-3

1. The doubling rate appeared to be short.
2. **Homologous recombination is the preferred mechanism of DNA repair in yeast.**
3. Often, jewel weed **grows close to** poison ivy.
4. Upon heat activation, filament size increased, and the number of buds decreased. Both **of these changes** were only seen for cytokinin mutants.

Problem 3-4

Mangrove plantations attenuated tsunami-induced waves and protected shorelines against damage. On the shorelines most damaged by the great 2006 tsunami, the area of mangroves had been reduced by 26% through human activities. By conserving or replanting coastal mangroves and green-belts, communities can be buffered from future tsunami events. The same buffer could be fulfilled elsewhere by the conservation of dune ecosystems or green belts of other tree species.

Problem 3-5   Word Placement and Flow

Fleas often carry organisms that cause diseases. An example of an organism that is transmitted by fleas to humans is the plague *bacillus*. These *bacilli* migrate from the bite site to the lymph nodes. Enlarged lymph nodes are called "buboes," giving rise to the name "bubonic plague."

Problem 3-6   Word Placement and Flow

Thermophiles are microorganisms that grow at a temperature range between 45°C and 70°C. They are found in hot sulfur springs. Because thermophiles cannot grow at body temperature, they are not involved in infectious diseases of humans. However, it is unclear how they resist ele-vated temperatures.

After reading the last sentence, the reader expects to hear about possible mechanisms of resistance.

Problem 3-7

1. Transplant rejection **increased**.
2. Amyloid plaques **were measured** twice for each brain. (**We measured** …)
2. Our results showed that the vaccine **protected** the dogs.
3. To determine whether cells **migrate**, we dissected mouse brains.
4. Buffalo, elephant, and black rhino abundance all **declined** rapidly after 1977.

Problem 3-8

1. A few microorganisms such as *Mycobacterium tuberculosis* are resistant to phagocytic digestion. **This resistance** is one reason why tuberculosis is difficult to cure.
2. Results indicate that binding decreased about 4-fold when the temperature increased from 4°C to 17°C. (Fig. 1). **This temperature increase/decrease/change** suggests that the binding among the particles may be governed by interactions such as hydrogen bonds or van der Waals forces that weaken when temperature rises.
3. The breakthrough was achieved because new methods and concepts could be applied, which had been developed in nonlinear dynamics to describe the spontaneous formation of order in various disciplines. **This description** in turn was only possible because the underlying laws are universal at a certain abstract level.

Problem 3-9

1. The pathogenesis observed in other cells, such as circulating monocytes, may differ from **that in** endothelial cells.
2. Diabetes can be affected **by both** exercise and diet. OR Diabetes can be affected **both by** exercise and **by** diet.
3. **Mutant A showed the highest peak.**
4. The male dolphin was **larger than** any of the other animals.
5. 'A' stars are less numerous **than** their solar-type equivalents.

Chapter 4 Fundamentals of Scientific Writing, Part II—Compositions

Problem 4-1

(**1'**) **Salmon use different methods to find their way during their homeward journey in the fall.** (**1**) Salmon may use a magnetic or sun compass to orient themselves. (**2**) As described by Brown et al. (15), **they may also use olfactory cues learned as smelts** to find the river and tributary of their birth. (**3**) Salmon may reenter fresh water in spring, summer, or fall, but spawning occurs in the fall, and the life cycle of the salmon begins anew.

The topic (1') *sentence states the message of the paragraph, "the different methods salmon use to find their way during their homeward journey in the fall." The topic is the subject in every sentence, resulting in a consistent point of view throughout the paragraph. Sentence 2 is parallel to sentence 1.*

Problem 4-2

A possible way to write this paragraph is the following:

CANCER CELLS
Cancer cells, which are malignant tumor cells, differ from normal cells in three ways: (1) they dedifferentiate—for example, ciliated cells in the bronchi lose their cilia; (2) they can metastasize, meaning they can travel to other parts of body; and (3) they divide rapidly, growing new tumors, because they do not stick to each other as firmly as normal cells do.

Problem 4-3

The active ingredient in most chemical-based mosquito repellents is DEET (*N,N*-diethyl-meta-toluamide), which was developed by the U.S. military in the 1940s. Although recent research suggests that DEET products, used sparingly for brief periods, are relatively safe (2), DEET should be used with caution as it is absorbed readily into the skin. Common side effects to DEET-based products include rash, swelling, itching, and eye irritation, often due to over-application (4). Some research even points to toxic encephalopathy associated with use of DEET insect repellents (5).

DEET alternatives include eucalyptus oil containing cineol, 1:5 parts diluted garlic juice, soybean oil, neem oil, marigolds, Avon Skin-So-Soft bath oil mixed 1:1 with rubbing alcohol, and the juice and extract of Thai lemon grass (7, 8). True citronella has not been proven to be a very effective mosquito repellent (3).

Problem 4-4

A potential natural hazard is an increased rate of sea level rise related to polar ice sheet decay. Although variations in Earth's orbit around the sun are considered to be a primary external force driving glacial–interglacial cycles, modern polar ice sheets have become unstable within the natural range of interglacial climates due to continued climate warming. Anthropogenic climate warming will accelerate the natural process toward reduction in polar ice sheets. As a result, the sea level is projected to rise between 13 and 94 cm over the next 100 yr.

Controversial geologic evidence suggests that current polar ice sheets have been eliminated or greatly reduced during previous Pleistocene interglacials. For example, there is conflicting evidence for warmer conditions and higher sea level during marine isotope stage 11. During marine isotope stage 11, the West Antarctic ice sheet and the Greenland ice sheet were absent or greatly reduced. The warm marine isotope stage 11 interglacial climate, during which sea levels were as high as or above modern sea levels, lasted for 25 to 30 ky. Based on microfossil and isotopic data from marine sediments of the Cariaco Basin, the global sea level was 10 to 20 m higher than today during marine isotope stage 11. During a presumably very warm interglacial about 400 ka during marine isotope stage 11, sea levels may have been more than 20 m higher than today.

Problem 4-5

*The paragraph can be condensed by only stating the most important informa-
tion and omitting all other irrelevant results.*

Our results indicate that only diamond covered with chemically bound
hydrogen shows a pronounced conductivity at temperatures between 5°C
and 25°C.

(21 words)

Chapter 5 Date, Figures, and Tables

Problem 5-1

1. Table or figure/map, depending if actual numbers are important or
    trends are meant to be displayed.
2. Table and/or figure/map—Table would provide precise data, but a map
    may show more visually the location of occurrence.
3. Table or graphs. Omit rat photo as it only shows a control rat.
4. Show the electron micrograph—it presents new evidence. The roent-
    genogram does not present new evidence and therefore should not be
    shown.
5. Both—a schematic will help to visualize what is being described in the
    text.
6. Both—a schematic will help to visualize what is being described in the
    text.

Problem 5-2

A figure of the average values per time reading would best present the data.

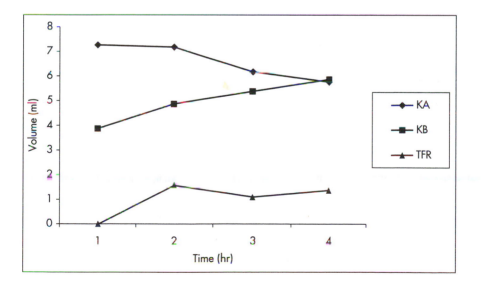

Problem 5-3

The best way to present this data would be as a bar graph.

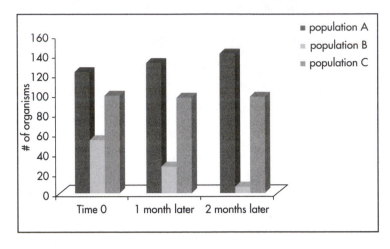

Problem 5-4

- Omit the title—it should go into the figure legend.
- Expand the *y*-axis scale so data points do not fall below the *x*-axis.
- Omit the grid lines.
- Tick marks should only face outwards.
- Remove the heavy frame around the graph.
- Replace "+" symbols with closed circles, squares, or triangles.

Problem 5-5

$n = 15$

Mean = 14.2

Variance = 3.2

Standard deviation = 1.8

Chapter 6 Laboratory Reports and Research Papers

Problem 6-1

Background	To remove cellular metabolic wastes from the body, mammals have an excretory system. To gain more insight into
Question/Purpose	such a system, we dissected a single kidney of a fetal pig. For this purpose, we cut from the lateral margin, leaving the
Experimental approach	ureter intact and connected to the renal pelvis. We found that the kidney is made up of three different regions internally: the outer cortex, the middle medulla (with the renal pyramids), and the innermost renal pelvis. Blood enters the kidney via the renal artery and is filtered by a maze of small tubules (nephrons) to remove water, ions, nitrogenous wastes, and other materials from the blood. The resulting urine then passed through the collecting ducts into the renal
Main findings	pelvis, which drains into the ureter and ultimately the urinary

Main findings	bladder for excretion. Filtered blood exits the kidney via the
Conclusion	renal vein. This study of the system shows how the mammal eliminates its fluid waste through urination.

Problem 6-2

In this Introduction, the unknown has not been stated, leaving the reader wondering what this study is really about and what it adds to the field.

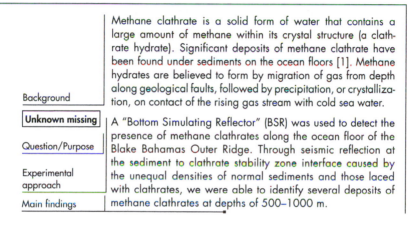

Background	Methane clathrate is a solid form of water that contains a large amount of methane within its crystal structure (a clathrate hydrate). Significant deposits of methane clathrate have been found under sediments on the ocean floors [1]. Methane hydrates are believed to form by migration of gas from depth along geological faults, followed by precipitation, or crystallization, on contact of the rising gas stream with cold sea water.
Unknown missing	A "Bottom Simulating Reflector" (BSR) was used to detect the presence of methane clathrates along the ocean floor of the
Question/Purpose	Blake Bahamas Outer Ridge. Through seismic reflection at the sediment to clathrate stability zone interface caused by
Experimental approach	the unequal densities of normal sediments and those laced with clathrates, we were able to identify several deposits of
Main findings	methane clathrates at depths of 500–1000 m.

Problem 6-3

The original paragraphs contain many experimental details that should not be placed into the Results section. These details need to be removed as shown:

To evaluate inhibitory effects of the selected molecules, enzymatic reactions were triggered by adding the substrate B and monitoring the absorbance of product. **We found that 6 out of 15 molecules showed appreciable inhibition at 10 μM (Figure 5). Three of those, A3, A6, and A7, exhibited more than 50% inhibition of the enzyme activity and were further diluted to find the minimal inhibitory concentration (MIC). Molecules A3 and A6 exhibited 30% and 45% inhibition of the enzyme activity, respectively, even at 1 μM. DMSO was found not to interfere with the enzyme.**

Problem 6-4

The paragraph lists only results and conclusions but is missing background/purpose information as well as the experimental approach:

Purpose missing	
Experimental approach missing	**To determine . . ., we . . .**
Results	We found that the H384A mutant reduced the k_{cat} value more than 3-fold. The apparent K_m values were increased 7-fold for Fru 6-P and 3.5-fold for PPi. The increase of the K_m values and

Interpretation	the reduction of the k_{cat} value of the H384A mutant suggest that the imidazole group of His384 is important for the binding stability as well as for catalytic efficiency of Fru 6-P and PPi substrates.

Problem 6-5

Version B is a better first paragraph. It signals the answer, which is followed by supporting evidence. Version A, in contrast, summarizes and repeats results without interpreting them.

Problem 6-6

Version A is a better concluding paragraph. The end is signaled clearly, the key findings are well summarized and interpreted, and the significance is indicated. Version B is simply listing limitations of the study, making for a very weak ending.

Chapter 10 Term Papers and Review Articles

Problem 10-1

The introduction of the review paper contains all necessary elements and is complete.

Background	Mitochondrial genomes differ greatly in size, structural organization, and expression both within and between and the kingdoms of eukaryotic organisms. The mitochondrial genomes of higher plants are much larger (200–2400 kb) and more complex than those of animals (14–42 kb), fungi (18–176 kb), and plastids (120–200 kb) (Refs. 1–4).
Unknown/Problem	Although there has been less molecular analysis of the plant mitochondrial genome structure in comparison with the equivalent animal or fungal genomes, the use of a variety of approaches—such as pulsed-field gel electrophoresis (PFGE), moving pictures (movies) during electrophoresis, restriction digestion by rare-cutting enzymes, two-dimensional gel electrophoresis (2DE), and electron microscopy
Topic statement	(EM)—has led to substantial recent progress. Here, the implication of these new studies on the understanding of *in vivo* organization and replication of plant mitochondrial
Overview	genomes is assessed.

(With permission from Elsevier)

Problem 10-2 **Review Abstract**

The abstract is that of a research article. Sentences 1 and 2 provide a short background. Sentence 3 signals the question, whereas sentence 4 states the experimental approach. Results are signaled in sentence 5. Sentence 6 provides the conclusion by stating an implication. No topic statement or overview sentence is provided.

Background	**(1)** Interleukin 1 (IL-1), a cytokine produced by macrophages and various other cell types, plays a major role in the immune response and in inflammatory reactions. **(2)** IL-1 has been shown to be cytotoxic for tumor cells. **(3)** To determine the effect of macrophage-derived factors on epithelial tumor cells, **(4)** we cultured human colon carcinoma cells (T84) and an intestinal epithelial cell line (IEC 18) with purified human IL-1. **(5)** Microscopic and photometric analysis indicated that IL-1 has a cytotoxic effect on colon cancer cells as well as cytotoxic and growth inhibitory effects on intestinal epithelial cells. **(6)** Because IL-1 is known to be released during inflammatory reactions, this factor may not only kill tumor cells but also affect normal intestinal cells and may play a role in inflammatory intestinal diseases.
Question/Purpose	
Experimental approach	
Results	
Answer/Conclusion/Implication	

Chapter 12 Oral Presentations

Problem 12-1

The slide shows a complex table taken directly from a published paper. Avoid using figures and tables directly as published in a paper, as most of the time you end up showing much more information than needed. If it is necessary to show a table, then reconstruct the table anew, and only show relevant portions on a slide. Avoid showing more than four columns and more than seven to eight rows. Listeners will not be able to absorb all this information at once, let alone read the small lettering of a reproduced figure or table. Also, consider highlighting values that are significantly different with a different colored font to draw the viewers' attention to the most important information on a slide.

Problem 12-2

Several problems exist for this slide: (a) The figures have been copied directly from a publication; (b) they lack a title and unnecessarily show a figure legend; (c) the writing is too small to read and the copies look fuzzy; (d) the font of the heading is a serif font—it is better to use a sans serif font, such as Arial; (e) the heading is uninformative; and (f) the slide looks crowded.

Problem 12-3

1. *This ending is very weak and shows that the presenter is not very confident. It is better to end just with a simple "Thank you."*
2. *Asking the audience to focus only on part of a slide and ignore the rest tells the listeners that the presenter has not prepared his or her talk very well and/or that the slide is overcrowded.*
3. *The speaker here is using written English instead of spoken English. This sentence sounds very awkward when used in speech.*
4. *Use of the passive voice and written style is very heavy and boring to listeners. Instead, say, for example, "We determined that. ..."*

Chapter 13 Posters

Problem 13-2

The sample poster panel can be much condensed. Bullet points will also make the main information more readily available. A revised poster panel is shown following.

Conclusion

- Recombinant bindin retains the ability to agglutinate eggs species-specifically.
- Species-specificity is contained in both amino- and carboxyl-terminal regions of *S. purpuratus* bindin.
- Repeated sequence elements in both amino- and carboxyl-terminal regions are different in bindins from different species.

Chapter 14 Research Proposals

Problem 14-1

	As the concentration of CO_2 in our atmosphere increases, so does the amount of CO_2 in the ocean's surface waters.
Background	Although the oceans are able to remove some of the increasing carbon from the atmosphere, just how much they are able
Problem	to remove has vital implications in determining the potential severity of global warming. To understand the global carbon
Objective	cycle, we propose to measure the CO_2 content of seawater over a 5-year period using high-precision methods for measuring total dissolved inorganic carbon (DIC), total alkalinity, and
Strategy	carbon and oxygen isotopes of CO_2. Results from this study will provide insight into CO_2 absorption by seawater and ulti-
Significance	mately increase our understanding of global warming.

Problem 14-2

Abstract

Cardiovirus is a common cause of gastroenteritis. For routine vaccination of Chinese infants, a new cardiovirus vaccine has been recommended. Here, we propose to evaluate the impact and cost-effectiveness of the Chinese cardiovirus vaccine program using a dynamic model of cardiovirus transmission. Findings from our analysis can be used to inform policy makers to prevent rotavirus infection.

Problem 14-3

This sample abstract contains all necessary elements.

Abstract

Background — The role that soils play in mediating global biogeochemical processes is a significant area of uncertainty in ecosystem ecology. One of the main reasons for this uncertainty is that we have a limited understanding of belowground microbial

Problem — community structure and how this structure is linked to soil processes. Building upon established theory in soil microbial ecology and ecosystem ecology, we predict that the structure of belowground microbial communities will be a key driver

hypothesis — of carbon and nutrient dynamics in terrestrial ecosystems. We propose to test and develop the established theories by

Overall objective — combining state-of-the-art DNA-based techniques for microbial community analysis together with stable isotope tracer tech-

Strategy — niques. By doing so, we expect to advance our conceptual and practical understanding of the fundamental linkages between soil microbial community structure and ecosystem-level carbon

Significance — and nutrient dynamics.

(Mark Bradford and Noah Frierer, proposal to private foundation)

Problem 14-4

This impact/significance statement focuses on the scientist's needs and wishes rather than on the funder's or on society. Most funders will not be interested in the number of papers that will be published. They would like to see how your efforts will advance the field and their cause.

Possible revision:

Funding from the Foundation will provide for a postdoctoral fellow and will lead to more insight into the fundamental principles of X, ultimately leading to better therapy and treatment options for Y.

Problem 14-5

The research proposal reads like an essay rather than a proposal. The text provides much background information and states the key questions to be answered but the overall objective is unclear. The objective is the most important statement of a proposal and needs to be included, even in boldface so the reader cannot miss it. In addition, in the text the specific aims are only alluded to but never stated outright. The largest section of the proposal, the experimental approach is completely absent. Thus, the reader comes away with interesting information about sea urchin gametes and their interactions but no idea about what the author is proposing to do and how.

A possible objective would be: To study the structure-function relationship of the sea urchin protein bindin and its receptor.

Specific aims could include:

1. *Express bindin as a recombinant protein in E. coli.*
2. *Prepare and study the activity and species specificity of a series of deletion mutants of bindin.*
3. *Isolate and identify bindin's receptor.*

Each of these aims should be followed by a detailed description of how the aim will be achieved experimentally and what expected outcomes would be.

Glossary of English Grammar Terms

Active Voice refers to the form of the verb used in relation to what the subject is doing. In English there are only two voices—passive and active. The *active voice* of a verb is the form of the verb used when the subject is the doer of the action.

Passive Voice: The manuscript was reviewed by the head of the department.

Active Voice: The head of the department reviewed the manuscript.

Adjective a word that modifies a noun or pronoun.

Example: The *young* birds accepted the bait readily.

Article a type of adjective which makes a noun specific or indefinite. In English there are three articles: the definite article *the* and the two indefinite articles *a* and *an*. In writing, an *article* is a brief nonfiction composition such as is commonly found in periodicals or journals.

Clause a group of words containing a subject and verb which forms part of a sentence.

Dependent/Subordinate Clause depends on the rest of the sentence for its meaning. It is usually introduced by a subordinating element such as a subordinating conjunction, transition word, or relative pronoun. It does not express a complete thought, so it does not stand alone. It must always be attached to a main clause that completes the meaning.

Noun a word that signifies a person, place, thing, idea, action, condition, or quality.

Object in the active voice, a noun or its equivalent that receives the action of the verb. In the passive voice, a noun or its equivalent that does the action of the verb.

273

Passive Voice the form of the verb used when the subject is being acted on rather than doing something.

Passive Voice: The manuscript was reviewed by the head of the department.

Active Voice: The head of the department reviewed the manuscript.

Person refers to the form of a word as it relates to the subject. In English, there are three persons:

First person refers to the speaker. The pronouns *I, me, myself, my, mine, we, us, ourselves, our,* and *ours* are first person.

Second person refers to the one being spoken to. The pronouns *you, yourself, your,* and *yours* are second person.

Third person refers to the one being spoken about. The pronouns *he, she, it, him, her, himself, herself, himself, his, her, hers, its, they, them, themselves, their,* and *theirs* are third person.

Phrase a group of words not containing a subject and its verb (e.g., *in this experiment, after the eruption*).

Plural *Plural* means "more than one." To show that a noun is plural, an *-s* or *-es* is normally added to the word. There are a few irregular plurals such as *men, children, women, oxen,* and a number of words taken directly from foreign languages such as *alumni* (plural of alumnus) or *media* (plural of medium).

Plural form of pronouns—pronouns that take the place of plural nouns such as *we, you,* and *they.*

Plural form of verbs—verbs that go with a plural subject.

Possessive Case the *possessive case* of a noun or pronoun shows ownership or association. Nearly all nouns and indefinite pronouns show possession by ending with an apostrophe plus an *s.*

Example: the organism's progeny

Preposition a word that relates a noun or pronoun (called the object of the preposition) to another word in the sentence. The preposition and the object of the preposition together with any modifiers of the object are known as a ***prepositional phrase.***

Prepositional Phrase a phrase beginning with a preposition and ending with a noun or pronoun. The phrase relates the noun or pronoun to the rest of the sentence. The noun or pronoun being related by the preposition is called the ***object of the preposition.***

Pronoun a word takes the place of a noun in a sentence.

Example: this, that, these, both, either

Sentence a group of words that expresses a thought. A sentence conveys a statement, question, exclamation, or command and must contain a verb and (usually) a subject. It starts with a capital letter and ends with a period (.), question mark (?), or exclamation mark (!).

Redundant means "needlessly repetitive."

Singular the form of a word representing or associated with one person, place, or thing. The term is normally used in contrast to plural. The singular form of a verb goes with a singular subject.

Subject the main noun (or equivalent) in a sentence about which something is said or who does something.

Tense the tense of a verb shows the time when an action or condition occurred (past, present, or future).

Verb the word or words that express action or say something about the condition of the subject.

Bibliography

Alley, Michael. *The craft of scientific writing*, 3rd ed. Springer, New York 1996.

Alley, Michael. *The craft of editing: A guide for managers, scientists and engineers.* Springer, New York, 2000.

Alley, Michael. *The craft of scientific presentations: Critical steps to succeed and critical errors to avoid.* Springer, New York, 2003.

Altman, Rick. *Why most PowerPoint presentations suck*, 1st ed. Harvest Books, 2007.

American Medical Association manual of style, 9th ed. Annette Flanigan et al. Lippincott, Williams & Wilkins 1997.

American National Standards Institute, Inc. *American national standard for the abbreviation of titles of periodicals.* American National Standards Institute, Inc., New York, 1969.

American National Standards Institute, Inc. *American national standard for the preparation of scientific papers for written or oral presentation.* American National Standards Institute, Inc., New York, 1979.

American National Standards Institute, Inc. *American national standard for writing abstracts.* American National Standards Institute, Inc., New York, 1979.

Atkinson, Cliff. *Beyond Bullet Points: Using Microsoft® Office PowerPoint® 2007 to create presentations that inform, motivate, and inspire.* Microsoft Press, 2007.

Booth, V. *Communicating in Science: Writing a scientific paper and speaking at scientific meetings*, 2nd ed. Cambridge University Press, Cambridge, 1993.

Browning, Beverly. *Perfect phrases for writing grant proposals*, 1st ed. McGraw-Hill, 2007.

Browner, Warren S. *Publishing and presenting clinical research.* Lippincott, Williams & Wilkins, 1999.

The Chicago Manual of Style. 15th ed. University of Chicago Press, Chicago, 2003.

Covey, Franklin. *Style guide for business and technical communication*. Franklin Covey Co., 1997.

Day, Robert A. *How to write and publish a scientific paper*, 5th ed. The Oryx Press, 1998.

Day, Robert A. and Gastel, B. *How to write and publish a scientific paper*, 6th ed. Greenwood, 2006.

Davis, Martha. *Scientific papers and presentations*. Academic Press, London, 2002.

Davis, M. and Fry, G. *Scientific papers and presentations*, 2nd ed. Academic Press, London, 2004.

Dodd, Janet S. *The ACS Style Guide: A manual for authors and editors*, 2nd ed. American Chemical Society, 1997.

Ebel, Hans. F., Bliefert, Claus, and Russey, William E. *The art of scientific writing: From student report to professional publications in chemistry and related fields*, 2nd ed. Wiley-VCH, 2004.

Feibelman, Peter J. *A PhD is not enough: A guide to survival in science*. Basic Books, 1993.

Foley, Stephen Merriam and Gordon, Joseph Wayne. *Conventions and choices. A brief book of style and usage*. D.C. Heath and Company, Toronto, 1986.

Fowler, H.W. *A dictionary of modern English usage*, 2nd ed. Oxford University Press, London, 1965.

Fowler, H.W. *A dictionary of modern English usage*. Wordsworth Editions Ltd., 1997.

Geever, Jane C. *The Foundation Center's guide to proposal writing*, 5th ed. Foundation Center, 2007.

Gerin, William. *Writing the NIH grant proposal: A step-by-step guide*. Sage Publications, Inc, 2006.

Gibaldi, Joseph. *MLA Handbook for writers of research papers*, 6th ed. The Modern Language Association of America, 2004

Glatzer, Jenna. *Outwitting writer's block and other problems of the pen*. The Lyon's Press, 2003.

Gopen, George D. *Expectations: Teaching writing from the readers perspective*. Pearson Longman 2004.

Gopen, George D. The science of scientific writing. *American Scientist*, Vol. 78, 1990.

Gopen, George D. and Swan, Judith A. The science of scientific writing. *American Scientist* Vol. 78, pp. 550–558, 1990. Also available on the web: https://www.americanscientist.org/issues/id.877, y. 0, no., content.true, page.1, css.print/issue.aspx, last accessed January 2009.

Greenbaum, Sidney. *The Oxford English Grammar*, 1st ed. Oxford University Press, New York, 1996.

Gustavi, B. *How to write and illustrate a scientific paper*. Cambridge University Press, Cambridge, 2003.

Hacker, Diana. *Rules for writers*, 5th ed. Bedford/St Martin's, 2004.

Hall, Mary S. and Howlett, Susan. *Getting funded: The complete guide to writing grant proposals*, 4th ed. Continuing Education Press, 2003.

Harris, Dianne. *The complete guide to writing effective & award-winning grants: Step-by-step instruction*. Atlantic Publishing Company, 2008.

Henson, Kenneth. *Grant writing in higher education: A step-by-step guide.* Allyn & Bacon, 2003.

Huth, E. J. *Writing and publishing in medicine,* 3rd ed. Lippincott, Williams & Wilkins, 1998.

Huth, E. J. *How to write and publish papers in the medical sciences.* Williams & Wilkins, 1990.

Iles, Robert L. and Volkland, Debra. *Guidebook to better medical writing,* revised ed. Iles Publications, 2003.

Illustrating science: Standards for publication. CBE Scientific Illustration Committee. Council of Biology Editors, Bethesda, MD, 1988.

International Committee of Medical Journal Editors. Uniform requirements for manuscripts submitted to biomedical journals: sample references. http://www.nlm.nih.gov/bsd/uniform_requirements.html, last accessed October 2012.

International Committee of Medical Journal Editors. Uniform requirements for manuscripts submitted to biomedical journals: writing and editing of biomedical publication. http://www.icmje.org/, last accessed October 2012.

American Medical Association manual of style: A guide for authors and editors, 10th ed. Oxford University Press, New York, 2007.

Katz, Michael Jay. *From research to manuscript.* Springer, 2006.

Knisely, Karin. *A student handbook for writing in biology,* 3rd ed. W. H. Freeman, 2009.

Knowles, Cynthia. *The first-time grant writers guide to success.* Corwin Press, 2002.

Korner, Ann M. *Guide to publishing a scientific paper.* Bioscript Press, 2004.

Lindsay, David. *A guide to scientific writing,* Longman, 2nd ed., 1995.

Lynch, Jack. *The English language: A user's guide,* Focus Publishing/R. Pullins Company, Newsburyport, MA, 2008.

Malmfors, B., Grossman, M., and Garnsworth, P. *Writing and presenting scientific papers.* Nottingham University Press, Nottingham, 2004.

Matthews, Janice R., Bowen, John M., and Matthews, Robert W. *Successful scientific writing.* Cambridge University Press, Cambridge, New York, 1996.

McMillan, Victoria E. *Writing papers in the biological sciences.* Bedford Books of St. Martin's Press, Boston, 1997.

Morgan, Scott and Whitener, Barrett. *Speaking about science: A manual for creating clear presentations,* 1st ed. Cambridge University Press, Cambridge, New York, 2006.

O'Connor Maeve. *Writing successfully in science.* Chapman and Hall, 1991.

Ogden, Thomas E. and Goldberg, Israel A. *Research proposals: A guide to success,* 3rd ed. Academic Press, 2002.

Peat, Jennifer and Barton, Belinda. *Medical statistics: A guide to data analysis and critical appraisal,* 1st ed. BMJ Books, 2005.

Peat, Jennifer, Elliott, Elizabeth, Baur, Louise, and Keena, Victoria. *Scientific writing: Easy when you know how,* 1st ed. BMJ Books, 2002.

Penrose, Ann M. and Katz, Stephen B. *Writing in the sciences: Exploring conventions of scientific discourse,* 2nd ed. Longman, 2004.

Perelman, Leslie C., Paradis, James, and Barrett, Edward. *The Mayfield handbook of technical and scientific writing.* Mayfield Publishing Company, 1997.

Pechenik, Ian A. *A short guide to writing about biology*, 7th ed. Longman, 2009.

Rogers, S. M. *Mastering scientific and medical writing: A self-help guide.* Springer, Berlin, 2006.

Rubens, Philip. *Science and technical writing: A manual of style.* Henry Holt and Company, Inc., New York, 1992.

Ryckman, W.G. *What do you mean by that? The art of speaking and writing clearly.* Dow Jones-Irwin, 1980.

Scientific style and format: The CBE Manual for authors, editors, and publishers, 6th ed. Council of Biology Editors, 1994.

Strunk, W., Jr. and White, E.B. *The elements of style*, 3rd ed. Macmillan, New York, 1979.

Sternberg, Robert. *The psychologist's companion: A guide to scientific writing for students and researchers*, 4th ed. Cambridge University Press, New York, 2003.

Style Manual Committee, Council of Biology Editors. *Scientific style and format: The CBE manual for authors, editors, and publishers*, 6th ed. Cambridge University Press, 1994.

Sullivan, K. D. and Eggleston, Merilee. *The McGraw-Hill desk reference for editors, writers, and proofreaders*, 1st ed. McGraw-Hill, 2006.

Taylor, Robert B. *Clinicians guide to medical writing*, 1st ed. Springer, 2004.

Teitel, Martin. *"Thank you for submitting your proposal": A foundation director reveals what happens next.* Emerson & Church, 2006.

The complete writing guide to NIH behavioral science grants, 1st ed. Editors: Lawrence M. Scheier, William L. Dewey. *Oxford University Press*, New York, 2007.

Thurman, Susan. *The everything grammar and style book.* Adams Media Corporation, Avon, MA, 2002.

University of Chicago Press Staff. *The Chicago manual of style*, 15th ed. University of Chicago Press, 2003.

Venolia, Jan. *Write right! A desktop digest of punctuation, grammar, and style*, 4th ed. Ten Speed Press, 2001.

Venolia, Jan. *Rewrite right!: Your guide to perfectly polished prose*, 2nd ed. Ten Speed Press, 2000.

Wason, Sara D. *Webster's new world grant writing handbook*, 1st ed. Webster's New World, 2004.

Williams, Joseph M. *Style: Lessons in clarity and grace*, 9th ed. Longman, 2006.

Williams, Joseph M. *Style: Ten lessons in clarity and grace.* Scott, Foresman & Co. 1988.

Yang, Jen Tsi. *An outline of scientific writing. for researchers of English as a foreign language.* World Scientific Publishing Company, 1995.

Yang, Otto O. *Guide to effective grant writing: How to write a successful NIH grant application.* 1st ed. Springer, 2005.

Young, Petey. *Writing and presenting in English: The Rosetta stone of science*, 1st ed. Elsevier Science, 2006.

Zeiger, Mimi. *Essentials of writing biomedical research papers*, 2nd ed. The McGraw-Hill Companies, 2000.

Credits

Problem 2-1 *Proceedings of the National Academy of Sciences* 105(5). Abrupt climate change and collapse of deep-sea ecosystems, pp. 1556–1560. Yasuhara, M., Cronin, T. M., deMenocal, P. B., Okahashi, H. and Linsley, B. K., Copyright 2008, National Academy of Sciences, U.S.A.

Example 6-24 Reprinted from *Developmental Biology* 156(1), Lopez, A., Miraglia, S. J. and Glabe, C. G. Structure/function analysis of the sea urchin sperm adhesive protein bindin, pp. 24–33, Copyright 1993, with permission from Elsevier.

Example 8-1 With permission from Tammy Wu, Yale University.

Example 8-2 With permission from Amanda Miller.

Example 10-3 From *Science* 5, 321. 2008. Rosenfeld, Daniel, Lohmann, Ulrike, Raga, Graciela B., O'Dowd, Colin D., Kulmala, Markku, Fuzzi, Sandro, Reissell, Anni, Andreae, Meinrat O. Flood or drought: How do aerosols affect precipitation?, p. 1309. Reprinted with permission from AAAS.

Example 10-5 Reprinted, with permission, from the *Annual Review of Ecology, Evolution and Systematic*, Volume 38. Aronson, R. B., Thatje, S., Clarke, A., Peck, L. S., Blake, D. B., Wilga, C. D., and Seibel, B. A. Climate change and invisibility of the Antarctic benthos, pp. 129–154. Copyright © 2007 by Annual Reviews www.annualreviews.org.

Example 10-6 Reprinted from *Science* 5, 321. 2008. Rosenfeld, Daniel, Lohmann, Ulrike, Raga, Graciela B., O'Dowd, Colin D., Kulmala, Markku, Fuzzi, Sandro, Reissell, Anni, Andreae, Meinrat O. Flood or drought: How do aerosols affect precipitation?, p. 1309. Reprinted with permission from AAAS.

Example 10-7 Reprinted, with permission, from the *Annual Review of Ecology, Evolution and Systematic*, Volume 38. Aronson, R. B., Thatje, S., Clarke, A., Peck, L. S., Blake, D. B., Wilga, C. D., and Seibel, B. A. Climate change and invisibility of the Antarctic benthos, pp. 129–154. Copyright 2007 by Annual Reviews www.annualreviews.org.

Index